SpringerBriefs in Electrical and Computer Engineering

Katsuaki Suganuma

Introduction to Printed Electronics

 Springer

Katsuaki Suganuma
Inst of Scientific & Industrial Research
Osaka University
Osaka, Japan

ISSN 2191-8112   ISSN 2191-8120 (electronic)
ISBN 978-1-4614-9624-3   ISBN 978-1-4614-9625-0 (eBook)
DOI 10.1007/978-1-4614-9625-0
Springer New York Heidelberg Dordrecht London

Library of Congress Control Number: 2013958230

Printed on acid-free paper

Springer is part of Springer Science+Business Media (www.springer.com)

Contents

Chapter 1
Introduction

1.1 Printing Technology in Electronics Manufacturing

Printed electronics (PE) has emerged as one of the key technologies not only for electronics but also for all kinds of electrically controlled machines and equipment. PE is a technology that merges electronics manufacturing and text/graphic printing. By this combination, one can manufacture high-quality electronic products that are thin, flexible, wearable, lightweight, of varying sizes, ultra-cost-effective, and environmentally friendly. All these features reflect the deep involvement of engineers in the development of PE technology.

This blended technology is, however, not new; it originated before the 1950s. Back then, some people started using printing to make circuits on printed wiring boards. In fact, there are reports on printing solutions for wiring in the 1950s. Figure 1.1 shows an example [1]. The researchers of Nippon Telegraph and Telephone found gravure printing was one of the promising printing methods for fine pitch accuracy. Nevertheless, printing did not emerge as the ultimate solution for wiring; the lithography of copper films bonded on glass-fiber-reinforced organic printed wiring boards came to be the standard technology for wiring board assembly. At the same time, ceramic substrate wiring boards processed by screen printing, though they had been in use in the production of ceramic packaging for one generation, is only a minor presence in the printed wiring board market, especially for server applications.

The next printing solution was displays. Shadow masks of TV cathode tubes had been fabricated by the combination of printing and etching. Fine pitch printing of original masks, down to 100 µm, was crucial for manufacturing fine display panels. Nowadays, flat panel displays, such as liquid crystal displays (LCDs) and plasma displays, are replacing cathode tube displays. LCDs in particular have become the main standard display technology. Such flat panel displays are also assembled with coating and printing processes.

On the other hand, ceramic passive components, such as capacitors, resistors, and antennas, required a fine printing process. Gravure printing and screen printing have been widely used for the production of ceramic passive components. Figure 1.2

K. Suganuma, *Introduction to Printed Electronics*, SpringerBriefs in Electrical and Computer Engineering 74, DOI 10.1007/978-1-4614-9625-0_1,
© Springer Science+Business Media New York 2014

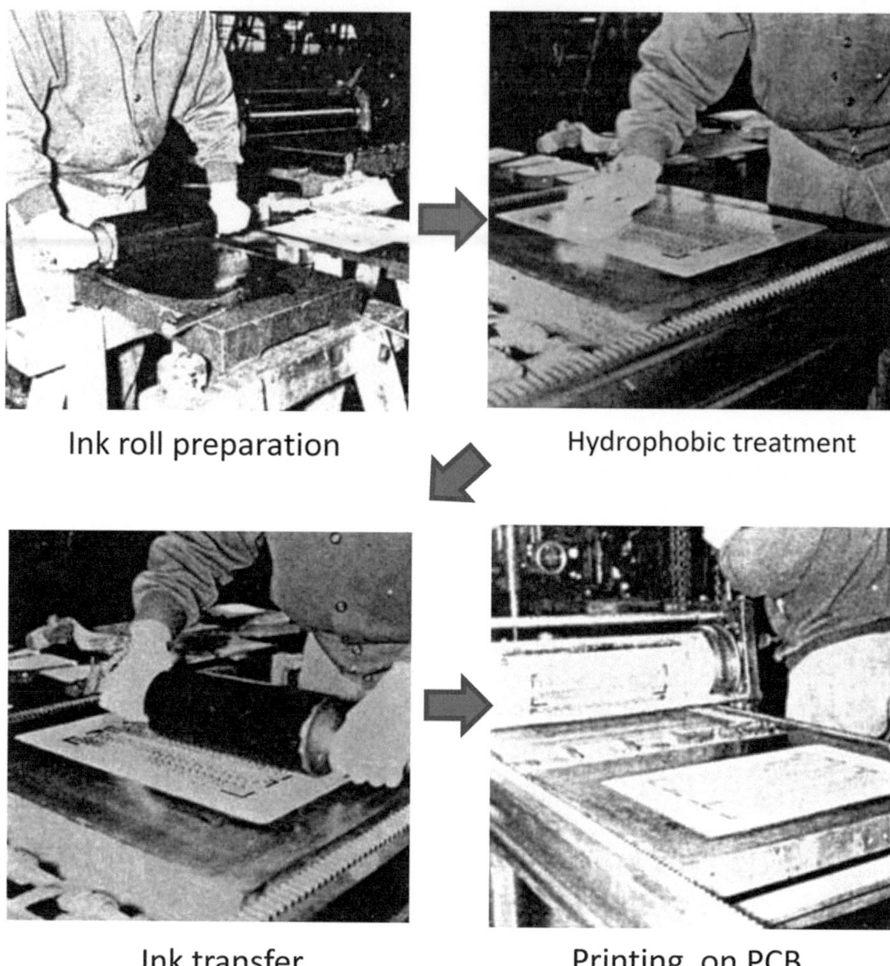

Fig. 1.1 Offset gravure printing of printed circuit board at Nippon Telegraph and Telephone, Tokyo, Japan [1]

shows a typical roll-to-roll screen printing of ceramic capacitors. Today, billions of tiny chip components, of which the smallest size is 0.4×0.2 mm, are manufactured continuously with Ni nanoparticle ink on ceramic green sheets.

Another example is solar cells. Solar cells based on Si technology also require screen printing and ink-jet printing in their manufacturing process. Finger grid lines and bus lines are formed by screen printing with Ag pastes containing glass flits (Fig. 1.3). The back plane contact is also formed by screen printing Al pastes. In addition, ink-jet printing is usually applied to form a doping line beneath the Ag lines on front planes.

Fig. 1.2 Fabrication of ceramic capacitor on substrate green sheet by roll-to-roll screen printing (Courtesy of Murata Manufacturing, Kyoto, Japan)

Fig. 1.3 Si solar panel and printed Ag paste grid and bas-bar

Most current electronics products possess surface-mount-type printed circuit boards that require wiring and soldering as one of the essential technologies. In soldering especially, the quality of screen printing of solder pastes plays a key role in the manufacture of small and high-functional products. Today, the smallest solder interconnection size comes in at below 100 µm. Figure 1.4 shows such a fine printed

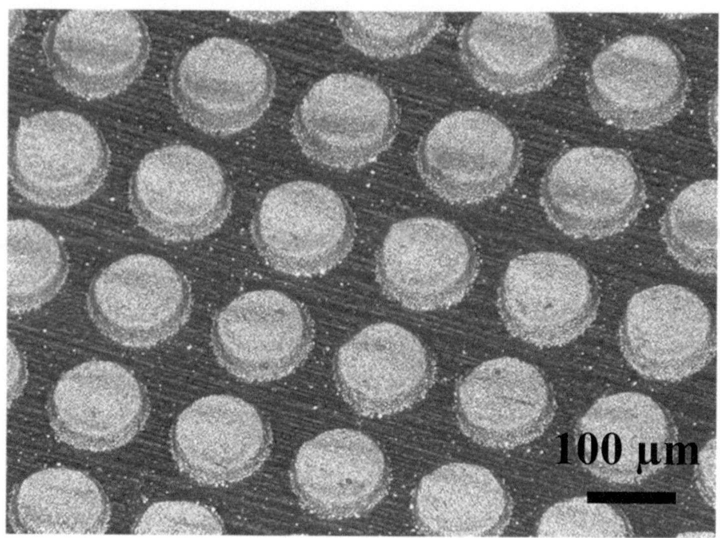

Fig. 1.4 Fine pitch solder bumps printed by screen printing (Courtesy of Harima Chemical, Hyogo, Japan)

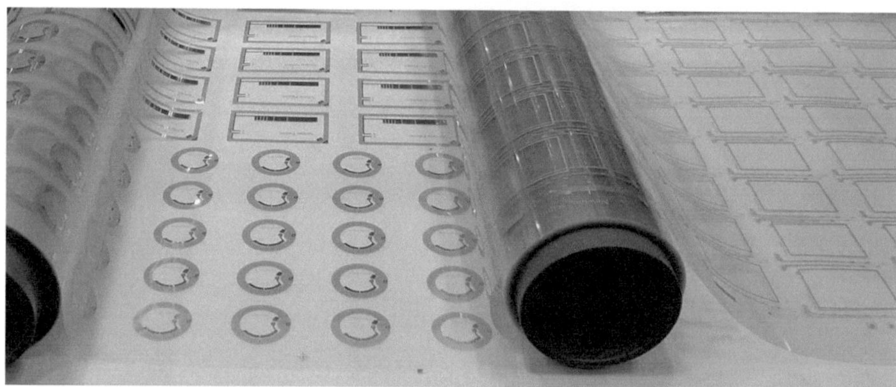

Fig. 1.5 RFID antenna and touch panel wiring with Ag-based conductive adhesive on PET film by rotary screen printing

solder paste on a printed circuit board. In some applications, conductive adhesives are used instead of solder pastes. For printed electronics, conductive adhesives, whether conventional micron-sized metallic flake pastes or newly developed nanoparticle pastes, are emerging as an essential interconnection technology that includes both wiring and bonding, which will be discussed in Chap. 6. Typical applications of conductive adhesives are the membranes of keyboards and touch panels (Fig. 1.5) and the antennas of radio-frequency identification (RFID) tags, which can be considered conventional printed electronics. Such products have been manufactured using an ultrafast printing method, i.e., rotary screen printing.

Thus, in recent decades, printing technology has grown with advances in electronic manufacturing technology, and there is great potential to significantly expand its field of use by combining this technology with the various advances in nanomaterials for electronics applications.

1.2   PE Technology and Its Benefits

As mentioned in the first section, PE is not a new idea that appeared in the twenty-first century; it grew gradually as part of electronics manufacturing in the twentieth century. In fact, many PE products already exist in the market. Nevertheless, great advances have been made in the past decade with the merging of print technology with nanomaterial technologies. The discovery of the basic nature of metallic, organic, and inorganic nanomaterials and their mechanisms and processes for synthesis, printing capabilities, electronic properties, and even evaluation methods have undergone tremendous advances thanks to the efforts of many scientists and engineers.

Let us now discuss some of the typical applications and major benefits of advanced PE technology. First, consider the cartoon in Fig. 1.6, which shows the PE products that are expected to make their way into our homes in the near future.

As can be seen, a large-screen TV hangs on the wall. This TV is lightweight, thin, and, perhaps, flexible. The TV panel itself is made of a self-light-emitting organic light-emitting diode (OLED) with an active matrix back plane made of

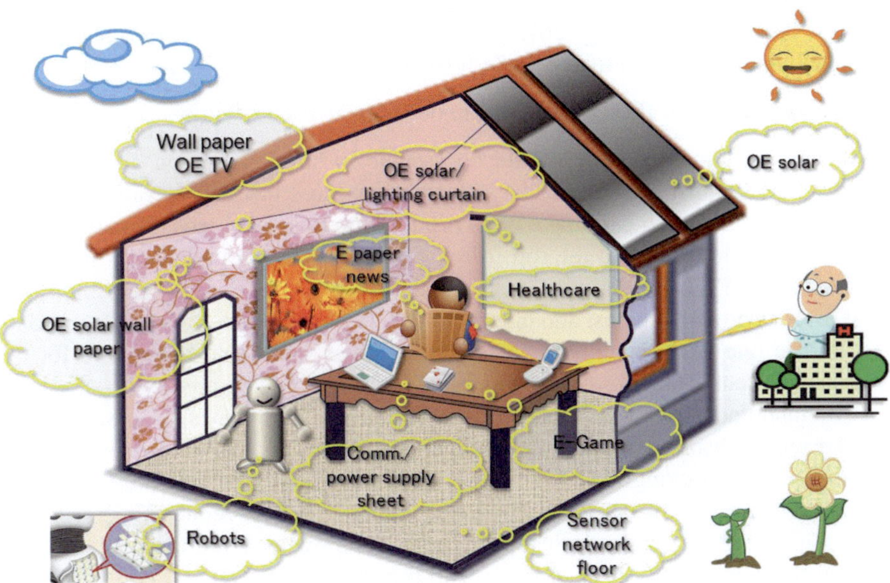

Fig. 1.6 PE technology in the near future

organic transistors with metallic nano ink circuits. The person sitting at the table is reading a newspaper, but it is not a simple paper. It is a actually a foldable display paper, perhaps like a future Kindle or iPad. Fresh content streams in throughout the day by wireless transmission over the Internet. The wall behind the TV with a pattern design is not a simple pattern but a dye-sensitized solar cell (DSSC) wall that recycles electrical energy from the lighting inside the house. The gadgets on the table—a smartphone, game cards, and notebook PC—are not merely sitting there but are being wirelessly charged by the communication sheet on the table and are also wirelessly connected to the Internet and an intranet. A robot is walking in the room. Because such humanoid robots must not injure people or pets or damage furniture, they must have a soft skin with a sensor network all over their bodies in every direction. The floor also has a sensor network beneath the carpet that senses any objects moving on the floor. The floor sensor network must also be soft. The curtain is not a simple cotton cloth. The outside face is an organic thin-film type of solar cell, and the inside face is an OLED lighting panel. The solar cell provides electricity to the internal lighting. The curtain itself works as a standalone flexible device. On the roof, of course, there is a solar cell module, possibly a thin-film inorganic type of module, such as a copper–indium–gallium–selenium (CIGS) one. Again, close inspection of the person sitting at the table reveals that he has some sort of device on his shoulder—a health monitor seal on his shirt. The seal monitors his temperature, blood pressure, pulse, sugar level, and other important health parameters. This sensor also works a standalone device and transmits health data to his doctor via cell phone.

Thus, a variety of PE products will be a regular feature of our lives in the near future and will provide valued comfort in our daily routines. These devices will not be noticed by people because they will be so thin, lightweight, form-fitting to walls, clothes, or even skin, energy efficient, and, above all, affordable. In other words, these will be the required features of PE technology.

The major benefits of PE technology can be summarized as follows:

1. It must be thin, lightweight, and be useable in large electronic devices—TV, solar, and lighting equipment can be larger than those made with conventional Si technology. Printing can make large products up to several tens of meters wide. Figure 1.7 shows one of the roll-to-roll screen printing examples of a RFID tag device on a PET (polyethylene terephthalate) film.
2. It reduces production cost and takt time: nowadays, Si technology has reached its ultimate fine pitch resolution, 13 nm, and a huge investment is required for the establishment of the production foundry. There are considerable risks associated with manufacturing short-lifetime products like cellular phones, tablets, and PCs. The most advanced semiconductor foundry cannot be maintained by a single enterprise even though it is very large one. Printing production requires less than approximately 1/10–1/100 the investment, and takt time is reduced considerably.

   Figure 1.8 shows the typical production of a printed semiconductor in a roll-to-roll process. Only four printers with pre- and post-treatment equipment are needed, just like a full-color gravure printing of graphic products. At the first printer, source and drain conductors are printed on a film. At the second printer,

Fig. 1.7 Screen printing of roll-to-roll devices on PET film (Courtesy of Tokai Seiki, Shiga, Japan)

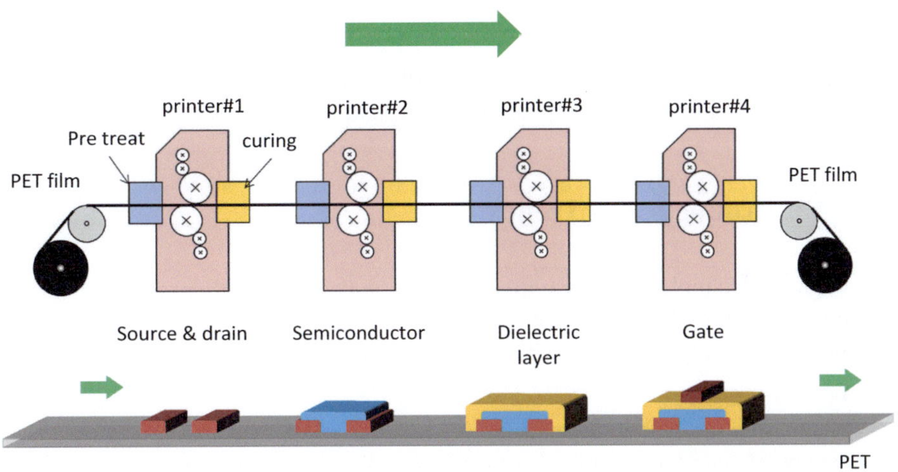

Fig. 1.8 Only four printers are needed to make a transistor

a semiconductor layer is printed on them. At the third printer, a dielectric layer is formed on them, and a gate electrode is then printed at the fourth printer. At each printer, pretreatment of the surface and after-curing are done in a short takt time. Because of the fast roll-to-roll printing speed, the takt time is much shorter than that for Si semiconductor manufacturing.

3. True wearableness: there is considerable demand for wearable devices. Conventional "wearable" devices are easily recognized due to their size, heavy

weight, stiffness, and fast power consumption. Truly wearable devices must be lightweight, thin, and comfortable and must power themselves.

4. All products need to be smart: all products, even pencils, will be equipped with some sort of intelligent device capable of communicating with the outside world wirelessly. Si dyes cannot be used in most of them because of silicon's high cost. Devices must be thin, tiny, lightweight, and inexpensive and have a self-contained energy supply.

5. Cheap devices are in high demand in developing countries: the Earth's population will exceed nine billion by 2050. Nowadays, populations are increasing only in developing countries. In many cases, they require affordable and renewable energy supplies.

6. PE technology is eco-friendly: all electronic devices must be environmentally friendly. Eco-friendly means free of toxins and rare earth elements and requiring low energy, both in the manufacturing process and in operation. There is also a strong demand for eco-friendly manufacturing processes. The reduction of solid and liquid waste in manufacturing is a key feature of PE technology, as is the fact that it uses less energy. In the conventional production of electronic equipment, complex processes, including lithograph technology, are required in the production of both Si devices and printed circuit boards. Much solid and liquid waste has been discarded in the environment. Printing does not require etching and so generates much less waste.

Thus, there are many reasons to explore PE technology not only to replace conventional electronics production but also to expand its applications and markets.

1.3 PE Products and Trends

Although PE technology is currently available on the market, there remains a huge potential market that will require much time and effort to develop. These market products can be categorized into the following groups:

- Lighting (OLED)
- Organic/inorganic photovoltaics
- Displays (front planes such as, for example, OLED, e-paper, and electrochromic and their active matrix back plane)
- Integrated smart systems (RFID, sports fitness/healthcare devices, smart cards, sensors, and smart textiles)
- Electronics and components (memories, antennas, batteries, wiring and interconnects, and other components)

Each device category has a potentially huge market. For instance, the worldwide new lighting market, which we will assume is equivalent to the current market, can grow by up to $150 billion in 2020, which is comparable to global annual sales of TVs [1]. In what follows, the features of some typical PE products and their current status, including their as-yet-unresolved issues, are summarized.

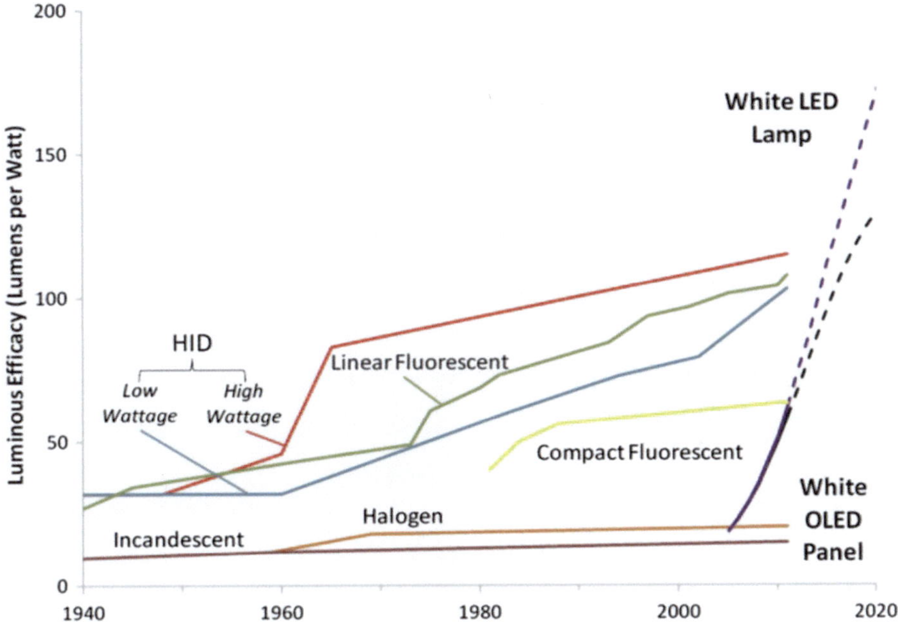

Fig. 1.9 Historical and predicted efficiency of light source [2]

1.3.1   Lighting

Since the two major conventional lighting systems possess fatal drawbacks, i.e., incandescent light bulbs consume a lot of electricity and fluorescent light tubes contain toxic elements, including mercury, which is banned by the Restriction of Hazardous Substances Directive 2002/95/EC of the European Union, it is urgent to replace them with certain environmentally friendly lighting systems. In Japan, for instance, the government released its assessment on lighting systems for the next two decades. All lighting units will be replaced with LED/OLED or other environmentally friendly lighting by 2020 and even in stockyards by 2030. This will reduce CO_2 emissions by approximately 25 % to 2012 levels. To achieve this, substantial improvements in lighting efficiency will be mandatory. Figure 1.9 shows the historical and predicted efficiency of light sources [2, 3]. The performance of halogen-incandescent, fluorescent, and high-intensity discharge light sources took 70 years to attain their current levels. In contrast, LEDs and OLEDs have experienced sharp efficiency increases due to extensive research progress, further increasing market penetration of LED/OLED lighting.

OLED lighting has a simple thin-film multilayer structure (Fig. 1.10). The feature that most distinguishes it from inorganic LEDs for application in lighting is its thin form factor. OLEDs produce light at relatively low intensity that spreads over large areas, while inorganic LEDs are more compact point sources or point arrays.

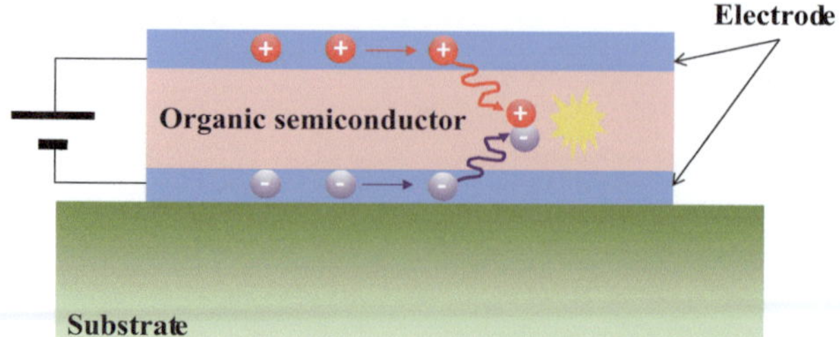

Fig. 1.10 Simple structure of OLED lighting

Fig. 1.11 Gravure printed
OLED lighting (Courtesy
of OLLA Project)

Figure 1.11 shows an example of a printed OLED panel. While the performance of commercially available OLED panels has not yet met lumen output or cost targets, considerable progress has been made in recent years. As indicated earlier, OLED lighting is expected to offer extraordinary potential with an efficiency that will substantially exceed that of traditional incandescent lighting sources. Thus, the differentiation with inorganic LED lighting will certainly play a key for the extension and success of OLED lighting technology.

1.3.2   Organic/Inorganic Photovoltaics

There are three major types of printed photovoltaic cell structures (Fig. 1.12). Dye-sensitized solar cells (DSSCs) have been on the market since 2012. Figure 1.13 shows the first mass-produced DSSC attached to an Apple (Cupertino, CA) iPad keyboard. DSSC can change its color by changing the thickness of the TiO_2 layer inside.

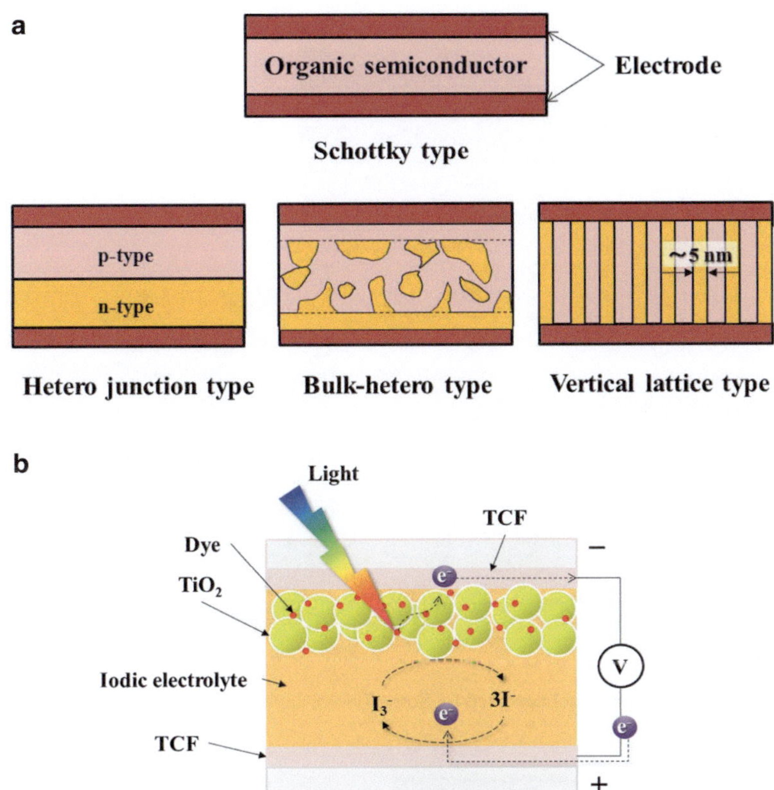

Fig. 1.12 Typical structure of (**a**) organic thin-film solar and (**b**) dye-sensitized solar cell

Fig. 1.13 iPad keyboard equipped with DSSC solar cell

Fig. 1.14 DSSC window panel designed by Sony, Tokyo, Japan

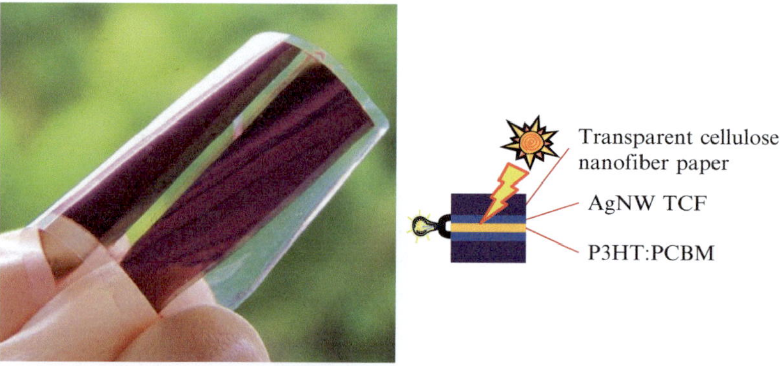

Fig. 1.15 Organic thin-film solar cell with cellulose nanofiber transparent paper [4] (Courtesy of Prof. M. Nogi, Osaka University, Osaka, Japan)

This simple technique can provide drawing and illustration to solar cells. Figure 1.14 shows an example of a designed window panel.

The second photovoltaic technology is an organic semiconductor. An organic thin-film solar cell has a layer structure similar to that of OLED lighting, as shown in Fig. 1.10. Figure 1.15 shows an example of a printed thin-film solar cell on a

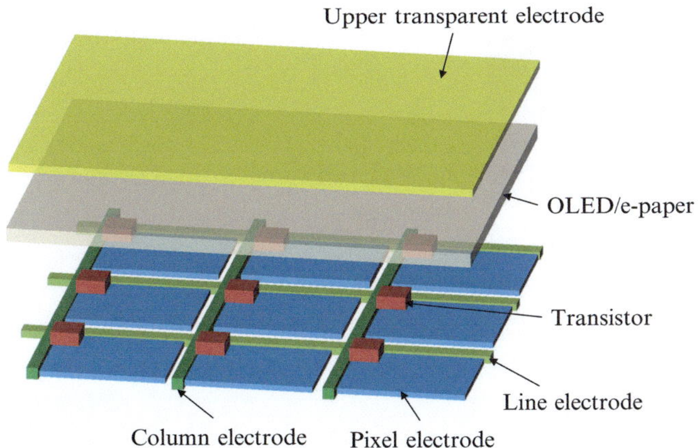

Fig. 1.16 Basic structure of active matrix driving flat panel display. In the case of an OLED active-matrix display, the OLED layer becomes a matrix of RGB pixels

transparent paper substrate [4]. It has a silver nanowire transparent conductive film as the top layer. The efficiency of light–electricity conversion is approximately 3 %. Due to the limited electron and hole transportation speed inside a molecule and the distance between organic molecules, the nanostructure modifications shown in Fig.1.12 are proposed to increase efficiency.

The inorganic semiconductor is the third photovoltaic technology. CIGS cells can provide much higher efficiency than DSSCs and organic thin-film solar cells. The commercial CIGS cells fabricated by a vapor process exceed by approximately 15 % in sunlight engineering efficiency, whereas DSSCs and organic thin-film-type cells are typically less than 5 %. The problem with CIGS cells is their high-temperature process; usually, a sintering temperature greater than 400 °C is required. It is expected that silicon nanoparticle or silicon compound inks will be used in the manufacture of printed photovoltaics, although they have the same problem.

1.3.3 Displays

OLED display technology is widely accepted as the most likely replacement for the cathode-ray tube and LCD. OLED displays offer several significant advantages over both technologies, such as enhanced clarity, a thinner, lighter weight design, and low energy consumption. In addition, if glass substrates and indium tin oxide electrodes can be replaced by flexible polymer or metallic substrates and electrodes, displays can be flexible and robust without being fragile. Moreover, organic materials can be processed into large-area thin films using simple and inexpensive printing technology. A basic structure of displays consists of a front panel and a passive-matrix or active-matrix TFT (Thin film transistor) back panel. An AM-TFT back panel is illustrated in Fig. 1.16.

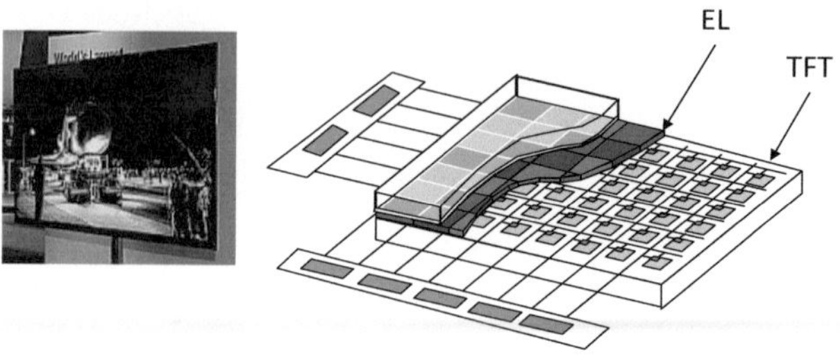

Fig. 1.17 World's first 4K printed OLED 55 in. display (Courtesy of Panasonic, Osaka, Japan)

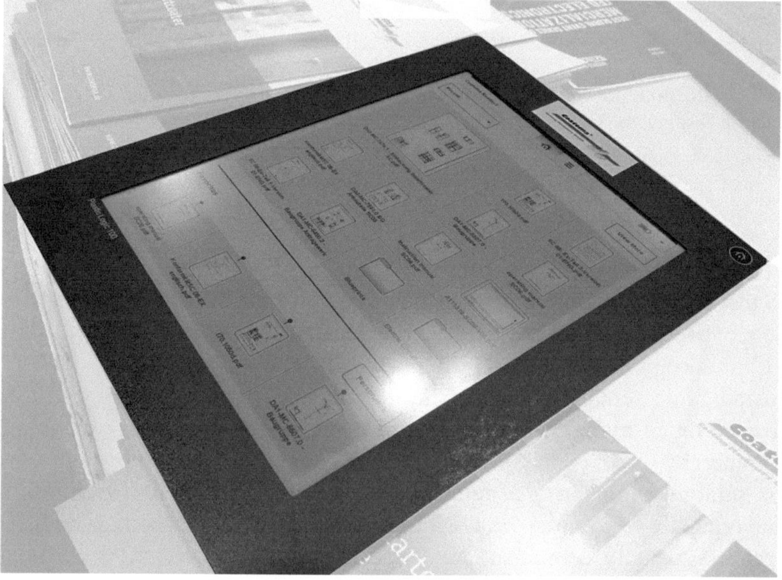

Fig. 1.18 A4-sized e-book with a printed back plane (Courtesy of Plastic Logic, Cambridge, UK)

At the CES (Consumer Electronics Show) 2013, Panasonic exhibited a 55 in. OLED display fabricated using PE technology, as shown in Fig. 1.17. This has 4K fine resolution. Other choices exist for a front display panel. E-paper is widely used for mobile displays such as e-books. E-paper display is voltage driven and has excellent memory effects, which makes for a long battery life, while OLED displays are current driven. Figure 1.18 shows the world's first A4 size e-book fabricated using PE technology released by Plastic Logic in 2011. Since this display has no glass substrate, it cannot be broken like LCDs when it is dropped from a height of 1 m.

Nevertheless, one must note the difficulty related to manufacturing displays. Display production requires precise pixel control of micron-level accuracy. This makes the mass production of large-area displays by printing very difficult. In addition, since reductions in the cost of conventional LCDs have occurred very

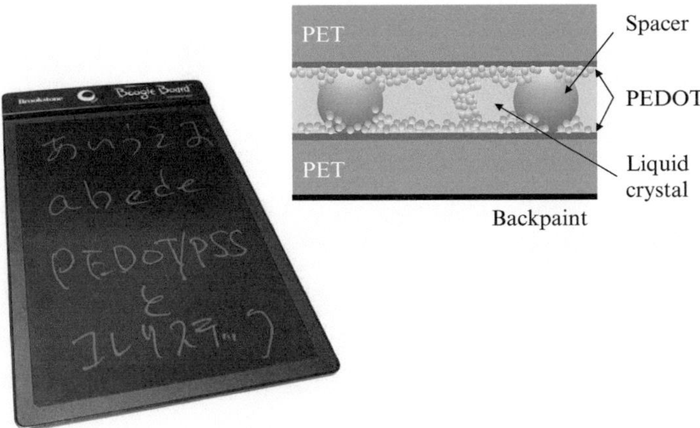

Fig. 1.19 Printed memo pad (Courtesy of Kent Display, Kent, OH)

rapidly and concomitantly with quality improvements, PE technology cannot simply be applied to the manufacture of displays for the replacement of LCDs.

There is an interesting example in display application. Figure 1.19 shows a memo-pad device made using PE technology by Kent Display [5]. Its simple structure consists of a cholesteric liquid crystal sandwiched between two PET films with a PEDOT/PSS (Poly(3,4-ethylenedioxythiophene) poly(styrenesulfonate)) electrode with no AM-TFT. This kind of simple breakthrough will be required before fine pitch displays find wide application.

1.3.4   Integrated Smart Systems

Integrated smart systems, such as RFIDs, smart cards, sports fitness/ healthcare devices, smart textiles, and various sensors, have very promising applications in PE technology. In fact, antennas of RFIDs and smart tags have been mass produced using screen printing with conductive adhesive paste, as shown in Fig. 1.5. Food traceability and medication control are two growing markets (Fig. 1.20). Although memory and oscillator components must be printed to expand the RFID market, printable high-speed transistors whose carrier mobility exceeds a few tens of square centimeters per volt per second will be necessary. Though recent advancements in semiconductors are truly amazing, as mentioned in Chap. 4, further development will likely require more time.

Sports fitness/wellness and healthcare applications require real wearable devices that are lightweight, thin, and conformal so that one is unaware of wearing the device. The devices must work as standalones either passively or actively and can communicate wirelessly outside the network system. Figure 1.21 shows a healthcare patch device made by MC10 that senses temperature, blood pressure, and glucose [6].

Fig. 1.20 RFID tags for medicine and food traceability management (Courtesy of PolyIC, Fuerth, Germany)

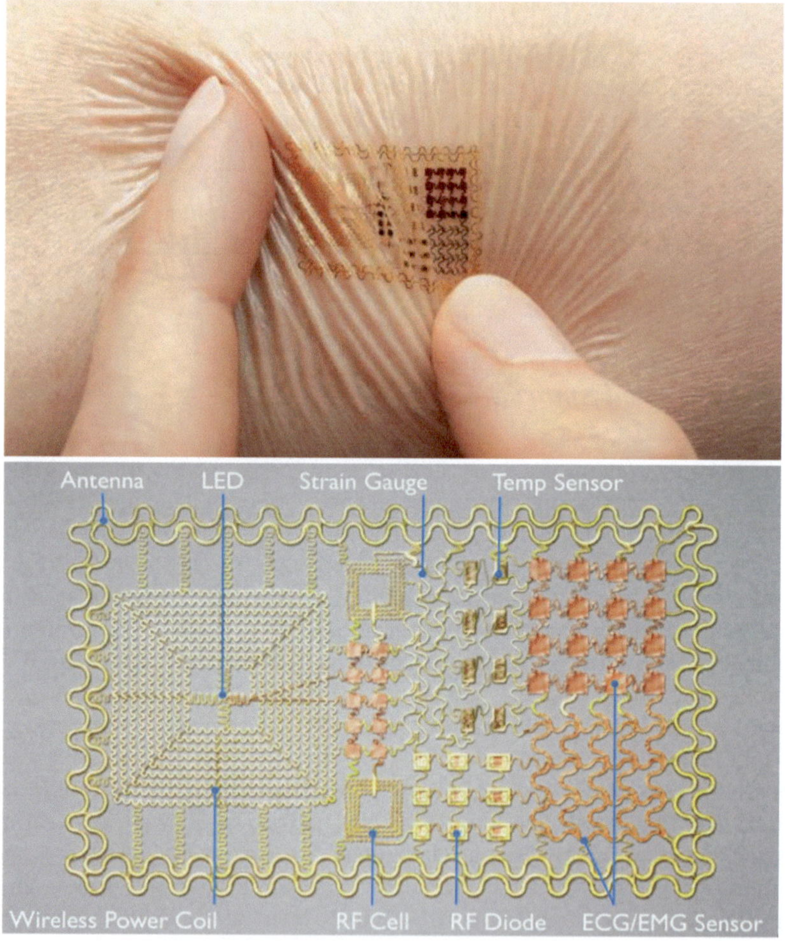

Fig. 1.21 Conformal electronics (Courtesy of MC10, Cambridge, MA)

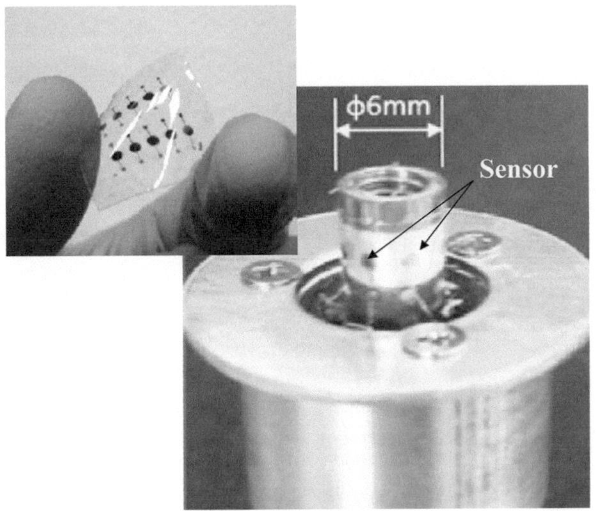

Fig. 1.22 Printed organic pyroelectric movement sensor on flexible substrate (Courtesy of Sensor-and-Works, Hyogo, Japan)

Movement sensors, mostly pyroelectric sensors, are becoming more popular with increasing security awareness in daily life as well as with the growing need to conserve energy. Currently, ceramic pyroelectric sensors are the main sensing devices. The flexibility of pyroelectric sensors also facilitates the use of printed sensors on flexible films. Figure 1.22 shows an example of an organic pyroelectric sensor screen printed on a flexible polymer film [7].

Various types of sensing devices can be fabricated using PE technology, for example, environmental sensors (temperature, humidity, gas concentration, ion concentration), biosensors (glucose, blood pressure, DNA), pressure sensors (floor mat sensor, touch sensor, explosion sensor), and light sensors. Many sensing devices have a simple semiconductor structure, as illustrated in Fig. 1.23. A sensing transistor layer is placed in between two source and drain electrodes. This can be a simple printabel TFT structure. A certain transistor material activated by a target species is printed between a source and drain electrode. Figure 1.24 shows an ammonia gas sensor with polyaniline interdigitated electrodes [8]. This sensor was fabricated by the inkjet printing of polyaniline nanoparticle films with Ag-based conducting interdigitated electrode arrays on a PET film. The sensor was further combined with a heater foil for operation at a range of temperatures. The sensor was found to have a stable logarithmic response to ammonia in a range of 1–100 ppm.

As sensing devices, wearable healthcare monitoring systems that monitor physiological events are also of considerable interest for healthcare, sports/fitness, and defense applications. While clothing-integrated electrochemical sensors hold considerable promise, such noninvasive textile-based sensing requires proper attention to key challenges, such as sample delivery to the electrode surface, sensor calibration, and robust interconnection.

Fig. 1.23 (**a**) Basic sensor structure and (**b**) source-drain current change at the detection event of of target species

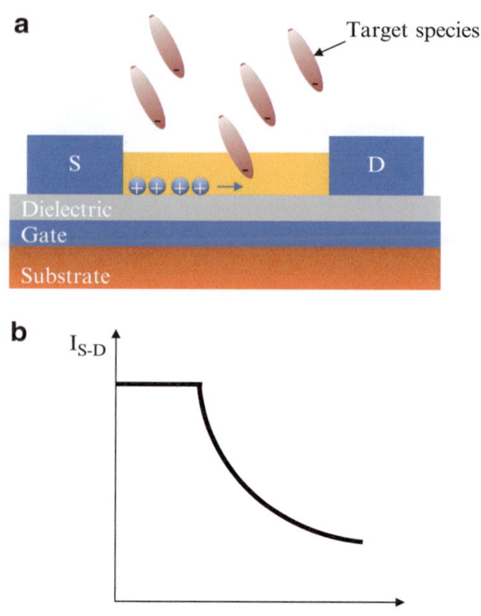

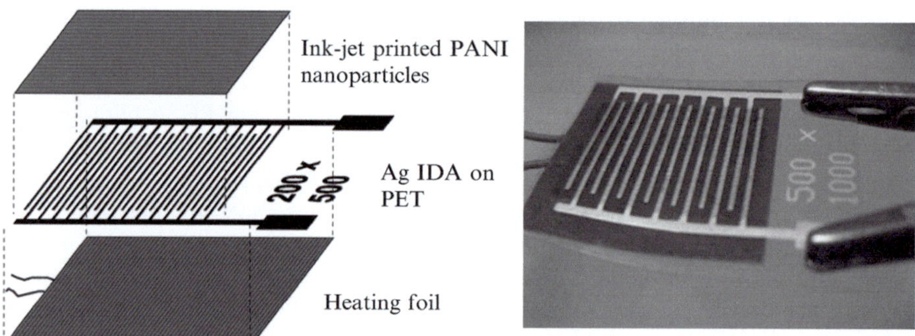

Fig. 1.24 Ammonia gas sensor with polyaniline interdigitated electrodes (nanoPANI-IDAs) [6]

1.3.5 *Other Electronics and Components*

Many other devices can be printed such as memories, antennas, batteries, touch panel interfaces, and wiring/interconnects. These devices are seeing increasing use and will be popular in the near future. Most goods will possess smart memory with wireless communication antennas.

Figure 1.25 shows an ink-jet-printed memory that has a basic passive structure consisting of a ferroelectric film sandwiched between two electrodes. When voltage is applied, the dielectric dipoles within the polymer layer align in one of two

Fig. 1.25 Inkjet printed
flexible memory (Courtesy
of Thin Film Electronics,
Oslo, Norway)

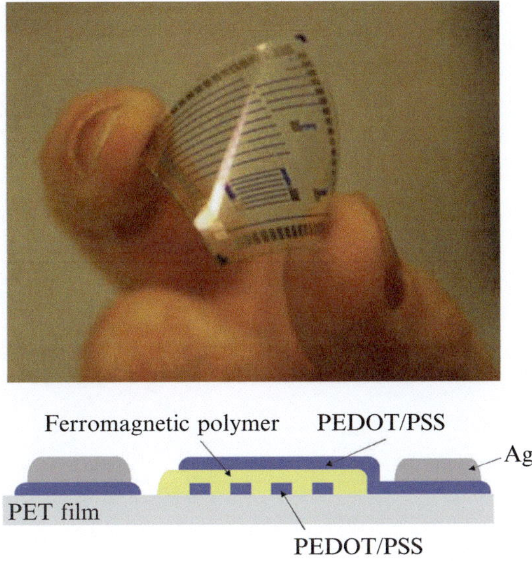

directions, depending on whether the voltage is applied to the top or bottom electrode. When the voltage is removed, the ferroelectric polymer layer maintains the
memory by pinning to the same state. Thus, it is a nonvolatile memory cell.

Printed antennas are currently made with conductive adhesives. However, their
resistivity is relatively high, 5×10^{-5} Ω cm, as compared with etched metallic foils
on the order of less than 5×10^{-6} Ω cm. To improve antenna properties, especially in
high frequency ranges up to the gigahertz range, metallic nano ink has significant
advantages. Figure 1.26 shows antennas on a PET film with Ag carboxylate ink [9].
This Ag carboxylate ink is cured at 80–100 °C, forming a mirror surface that facilitates high-frequency transmission. Metallic nanowires may also provide an effective antenna structure (Chap. 3).

Flexible wiring will expand the PE market. Epson was the first to develop a multilayer flexible tag circuit (Fig. 1.27). The entire structure, including the via-hole,
was processed using inkjet printing. Twenty layers were formed on the film, which
had a total thickness of 200 µm.

Transparent and flexible wiring has been made possible by PE technology.
Figure 1.28 shows an example of transparent wiring on a PET film made by screen
printing Ag nanowire inks (Chap. 3).

PE technology has also seen widespread application in batteries and capacitors.
The structure of batteries is very simple. Figure 1.29 shows commercial products as
cosmetic application.

Thus, the potential near-future market for PE is in the multibillion-dollar range.
As mentioned earlier in this chapter, PE applications are expanding into many
electronic products. Photovoltaics, flexible displays, lighting, textile electronics,
sensors, and other integrated smart systems are just some of the markets that will be
revolutionized by PE technology.

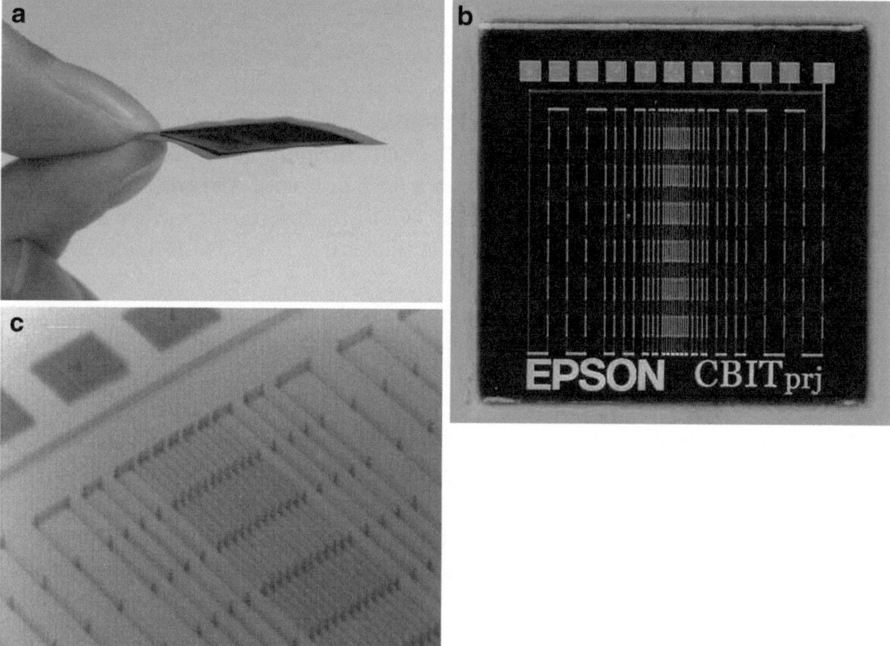

Fig. 1.27 Flexible 20 layers wiring printed circuit board (PCB) made with inkjet printing (Courtesy of EPSON, Nagano, Japan). (**a**) side view and (**b**) Top view. (**c**) is an X-ray transmission image showing via wiring inside the PCB.

Fig. 1.28 Flexible transparent seven-segment display formed on PET. Wiring was formed with AgNWs on PET film by photosintering. LEDs were mounted on PET film with low-temperature, curable, conductive adhesive. Sample was fabricated for demonstration in collaboration with Samsung Electronics (Seoul, Korea), Shows Denko (Tokyo, Japan), and Okuno Chemical Industries (Osaka, Japan)

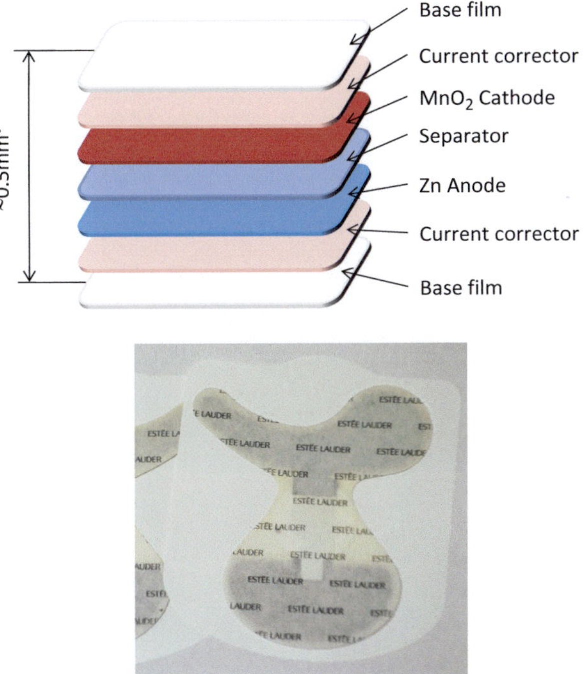

Fig. 1.29 Structure of printed primary ZnMnO$_2$ cell and application for cosmetic face patch

References

1. Nagatsu H (1959) Printed wiring board fabrication by offset gravure printing, Nippon Telegraph and Telephone Public Corporation Report, No. 45
2. T. Baumgarter, F. Wunderlich, D. Wee, A. Jaunich, Lighting the way: Perspectives on the global lighting market. Issue No. 3, McKinsey & Company, October 201, http://www.enlighten-initiative.org/portal/Home/tabid/56373/Default.aspx
3. Solid-state lighting research and development: multi-year program plan. US Department of Energy, April 2012
4. Tokuno T, Nogi M, Karakawa M, Jiu J, Aso Y, Suganuma K (2011) Fabrication of silver nanowire transparent electrodes at room temperature. Nano Res 4:1215–1222
5. Schneider T, Magyar G, Barua S (2008) A flexible touch-sensitive writing tablet. Dig Tech Pap SID Int Symp (Soc Inf Disp) 39(3):1840–1842
6. Kim D-H, Lu N, Ma R, Kim Y-S, Kim R-H, Wang S, Wu J, Won SM, Tao H, Islam A, Yu KJ, Kim T-I, Chowdhury R, Ying M, Xu L, Li M, Chung H-J, Keum H, McCormick M, Liu P, Zhang Y-W, Omenetto FG, Huang Y, Coleman T, Rogers JA (2011) Epidermal electronics. Science 12:838–843
7. Horie S, Ishida K (2012) Human motion sensing application of the organic ferroelectric material. Ceramics 47(10):797–801
8. Crowleya K, Morrina A, Hernandeza A, O'Malleya E, Whittenb PG, Wallace GG, Smytha MR, Killarda AJ (2008) Fabrication of an ammonia gas sensor using inkjet-printed polyaniline nanoparticles. Talanta 77(2):710–717
9. Hirose K, Kawazome M, Sekiguchi T, Hatamura M, Suganuma K (2012) Low temperature wiring technology with silver β-ketocarboxylate. IEICE Trans Electron J95-C(11):394–399

Chapter 2
Printing Technology

2.1 Printing Parameters

A wide range of printing methods have already been applied to conventional electronics fabrication. They include screen printing, inkjet printing, gravure printing, flexo printing, and offset printing. They are also applicable to many advanced PE products. Depending on the nature of the PE products, one must make a suitable choice regarding the of ink, substrate, designed device structure, pattern geometry, manufacturing speed, yield, quality, and production cost. The important printing parameters are as follows:

- Printing accuracy and resolution: display application for smart phone/tablets, among the finest applications today, requires fine patterning above 300 pixels per inch (ppi). A resolution of a few micrometers with ±5 μm position accuracy will be required. Multilayer printing accuracy is also a key factor.
- Uniformity from a few centimeters to more than 1 m in size area is required in combination with the designing ink composition and the drying process.
- Wetting control and interface formation: flatness within a few nanometers to several tens of nanometers is required for many OLED applications such as TV and lighting since the typical OLED layer thickness is less than 100 nm. Sharpness at pattern edges and bonding with substrates are strongly dependent on the underlayer (acceptance layer) material and its design.
- The compatibility of inks with printing components such as rollers, masks, doctor blades, and inkjet heads has a significant effect on yield and quality in mass production.
- Throughput and cost considerations: one of the great benefits of PE technology is its mass production at a reasonable cost. The high speed and high quality of printed patterns should be maintained for up to hundreds of printings.

Roll-to-roll printing is one of the active research areas in PE technology because it enables large-scale production by high-speed web handling. Roll-to-roll printing allows for large-scale production of such items as RFID antennas or keyboard

K. Suganuma, *Introduction to Printed Electronics*, SpringerBriefs in Electrical and Computer Engineering 74, DOI 10.1007/978-1-4614-9625-0_2,
© Springer Science+Business Media New York 2014

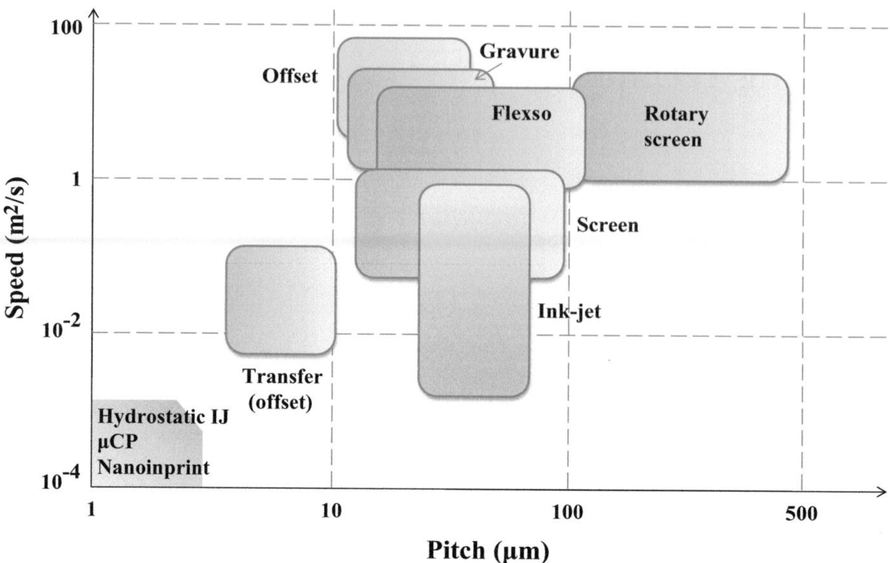

Fig. 2.1 Throughput vs. fine pitch comparison for various printing methods (Adapted from ref. [1] by author)

membranes. Nevertheless, the roll-to-roll process is not mature enough to be applied to many areas where PE technology is used since adjustments among materials, printing methods with suitable web handling, accurate positioning, and inspection methods with definitions of defect criteria have not yet been established. Sheet-fed production is still a major printing method for most PE products. To shift from sheet-fed printing to roll-to-roll printing will require time to develop the printing technologies with suitable parameters including materials development. Figure 2.1 compares the throughput to fine pitch resolution among various printing methods [1] (Table 2.1).

The choice of printing methods is sometimes a major issue before launching research projects or before building up production lines. There is no single selection for one application. There are certain suitable matchings between inks and printers. A substrate may play a role in this choice. Not only the viscosity/surface tension of the ink but the device structures and whether the device line/layer is thin or narrow will affect the pattern quality obtained.

The cross-section profile of a printed circuit or a device has a distinctive shape. Figure 2.2 shows typical wiring cross-sectional shapes formed by printing. As wiring or as a device, a square cross section as in Fig. 2.2a is desirable to obtain certain electronics properties. Unfortunately, this does not happen with PE technology except with high-viscosity inks such as in screen printing. In wiring by inkjet printing, a low-viscosity ink droplet lands on a substrate, so that its cross section sometimes exhibits a coffee-ring effect, as shown in Fig. 2.2c, depending on the viscosity of the ink, its wettability on a substrate, and the vaporization uniformity of the solvent.

Table 2.1 Feature comparison of printing methods

Printing method	Ink viscosity (cP)	Line width (µm)	Line thickness (µm)	Speed (m/min)	Other feature
Inkjet	10–20 (electrostatic inkjet: Approx. 1,000)	30–50 (electrostatic inkjet: Approx. 1)	Approx. 1	Slow (rotary screen: 10 m/s)	Surface tension: 20–40 dyn/cm On demand Noncontact
Offset	100–10,000	Approx. 10	Several –10	Middle—fast Approx. 1,000	
Gravure	100–1,000	10–50	Approx. 1	Fast Approx. 1,000	
Flexo	50–500	45–100	<1	Fast Approx. 500	
Screen	500–5,000	30–50	5–100	Middle Approx. 70	
Dispense	1,000–10^6	Approx. 10	50–100	Middle	Single stroke
µCP	–	Approx. 0.1	Approx. 1	Slow	
Nanoinprint	–	Approx. 0.01	Approx. 0.1	Slow	

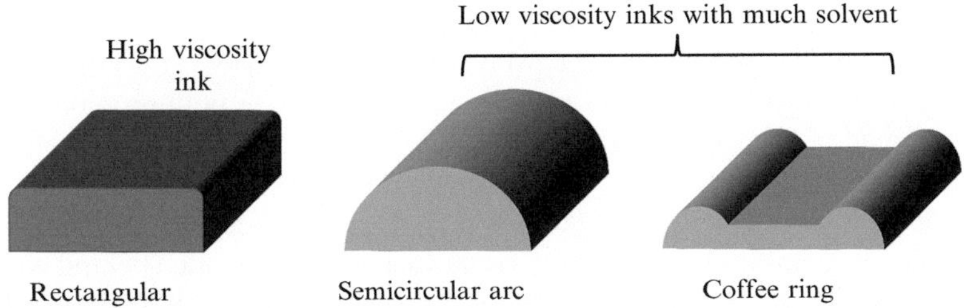

Fig. 2.2 Typical cross sections of printed patterns

This shape is not appropriate in most cases because many defects are likely to form in the concave central area. Thus, the droplet shape must somehow be made flat. At the very least, a semicylindrical shape, as in Fig. 2.2b, is desired.

The wetting ability of liquid on a solid substrate is measured by a simple sessile drop method, as shown in Fig. 2.3a. The wetting angle can be a good index for wetting. Where the wetting angle θ is larger than 90°, it is called nonwetting, while at less than 90°, it is known as wetting. The wetting phenomenon is governed by the surface energies of the liquid and of the substrate and the interface energy as expressed by the inset Young's equation. In drying patterned ink droplets, the segregation of solute content toward the outside edge of a droplet sometimes occurs, as

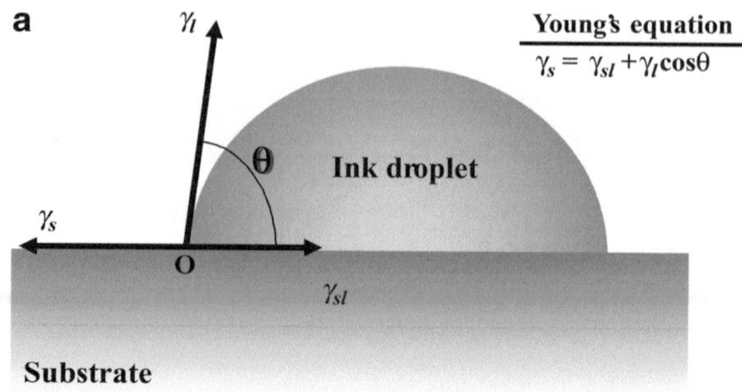

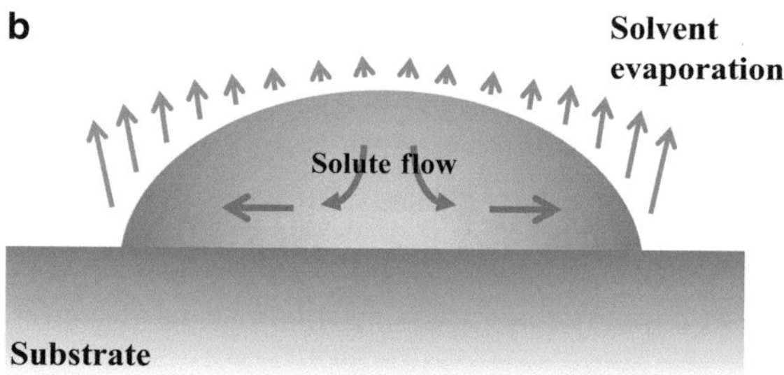

Fig. 2.3 (a) Wetting of a droplet on substrate. (b) Mechanism of coffee ring effect

shown in Fig. 2.3b, which is known as the coffee-ring effect. The coffee-ring effect must be avoided (see subsequent discussion).

To obtain the desired shapes for printed wires and devices, one must adopt a certain kind of wetting control on substrate faces, which can both promote and prevent the spread of printed inks. Basically, there are two ways to control wetting and spreading, which have been in use in the graphic printing industry for many years, i.e., chemical treatment and physical treatment. Figure 2.4 depicts these methods.

Making a low-/high-energy state of the surface is the basic idea behind chemical treatments. Table 2.2 summarizes the surface energy ranges for various types of polymer substrate. Thus, in addition to a substrate, the type of ink solvent is very important. Plasma cleaning of substrate surfaces usually creates a high-energy state on most surfaces, resulting in the promotion of wetting and spreading. CF_4 plasma treatment, in contrast, creates a fluorinated layer on the surface in a very low-energy state. Figure 2.5 shows a CF_4 treatment period on the contact angle of PEDOT/PSS (Poly(3,4-ethylenedioxythiophene) poly(styrenesulfonate)) droplets on substrates, a bank, and indium tin oxide (ITO). A CF_4 plasma treatment initially increases the contact angle very rapidly, and the increase slows down after 1 or 2 min. A selective hydrophobic treatment such as fluoridation can create a fine pitch pattern at widths of even less than 1 μm. Figure 2.6 shows an example of an extreme case of PEDOT/PSS fine patterning [2]. In this case, the first PEDOT line was inkjet printed and cured. Then, a CF_4 plasma treatment made the surface of PEDOT fluorinated low energy and the glass substrate high energy. Then the second PEDOT droplet flows off the low-energy first PEDOT surface, resulting in a very small gap formation between the two PEDOT lines.

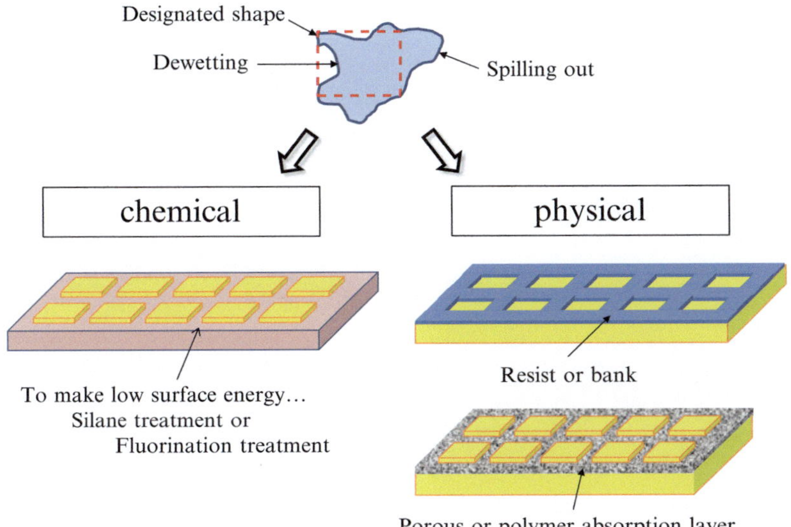

Fig. 2.4 Various wetting control methods

Table 2.2 Surface energies of various film substrates

Surface energy (dyn/cm^2)	Typical plastic	Properties
10–20	Silicone	Water repellent
	Fluorocarbon polymers	
20–35	Polyethylene	Hydrophobic
	Polypropylene	
35–50	Polyester	Polar
	Nylon	
	Epoxy	
	Acrylic resin	
	PET	
50–60	Polyvinyl alcohol	Hydrophilic
	Cellulose	
	Polyvinylpyrrolidone	

Creating a bank or an absorption layer is a reasonable method of forming accurate patterns that can be adjusted for many types of ink solvent. Bank formation, which is commonly used for display pixels, is limited in resolution by its printing method. Screen printing is usually used for bank formation where the bank width is larger than 30 μm. Ink-absorption-layer formation has been widely used in graphic printing. There are two different choices of absorption layer, a porous layer type and a polymer type. The combination of ink, especially a solvent, and substrate plays a key role in the successful control of wetting and absorption for both types of absorption layer. Figure 2.7 shows an example of inkjet-printed Ag nanoink line formation on a porous surface layer made with a silica sol-gel coating on a PET film.

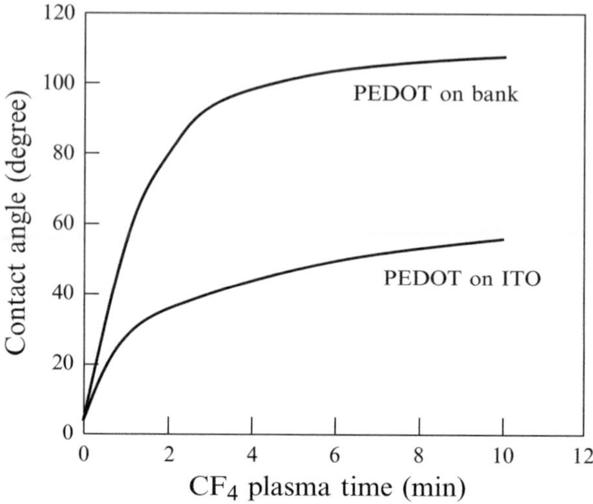

Fig. 2.5 Gas bombardment effect of PEDOT wetting (Courtesy of Dr. James Lee, CDT)

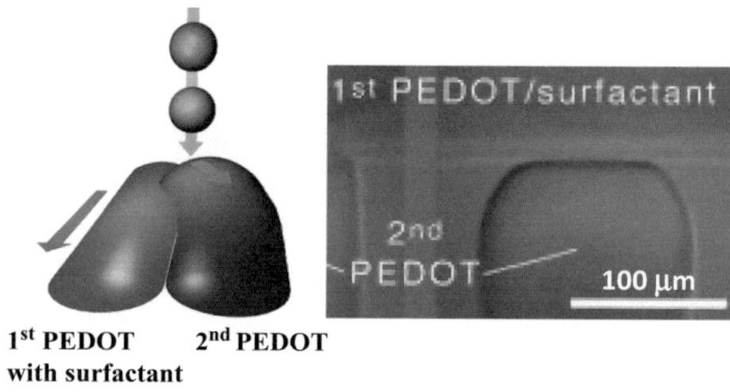

Fig. 2.6 Fine gap formation using CF$_4$ plasma wetting control [2]

The solvent was organic. The wetting of the Ag ink on the PET film was so good that ink spreads unexpectedly over the PET film. In contrast, ink wetting on the silica-coated PET film was precisely controlled, as expected.

Not only wetting of ink on substrates but a drying condition is important. For many printing methods such as inkjet, offset-gravure, and flexo, the viscosity of inks, the concentrations of metallic nanoparticle inks are very low, which means those inks contain a large amount of solvent. Because of the presence of much amount solvents, the solvent must be evaporated to achieve suitable solid tracks. Depending on the evaporation process, such inks often produce a coffee-ring effect, as mentioned earlier and shown in Fig. 2.3b, thereby unexpectedly resulting in high electrical resistance. Figure 2.8 shows the influence of the coffee-ring effect on the resistivity of wiring using Ag nanoparticle ink [3]. By changing the line width, the resistivity of the lines narrower than 300 μm is much greater than those of wider lines, of which resistivity is 5×10^{-6} Ω cm. The coffee-ring effect is caused by a convection flow from the center to the edge of droplets during the relatively slow

Fig. 2.7 Surface control effect of PET film substrate with silica sol-gel coating. Ag nanoparticle ink was inkjet printed on PET with or without silica coating

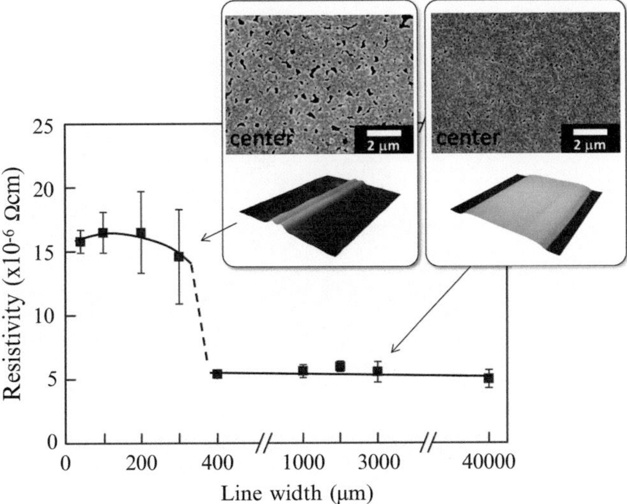

Fig. 2.8 Influence of line width on measured resistivity [3]; low-viscosity ink sometimes forms coffee ring pattern

evaporation of solvent. This coffee-ring effect can be overcome by reducing the solute flow in a drying ink droplet. The formation of absorption layers of ink vehicles on pristine polymer films is one of the effective methods that leads to the fabrication of convex-shaped lines without the coffee-ring effect, even if a low concentration of commercially available ink is used.

2.2   Screen Printing

Screen printing is one of the most common printing methods and has been used for many years in electronics manufacturing. The most distinct feature of screen printing compared with other printing methods is the high aspect ratio of printed objects. The usual thickness of a screen-printed image is in the range of several tens of microns, but, especially when a thick screen mesh is used, the thickness can exceed 100 μm with a single pass of printing, which cannot be obtained by any other printing method. For other methods such as inkjet or flexo printing, the typical thickness is less than 5 μm. Figure 2.9 shows a high-aspect-ratio screen-printed line example.

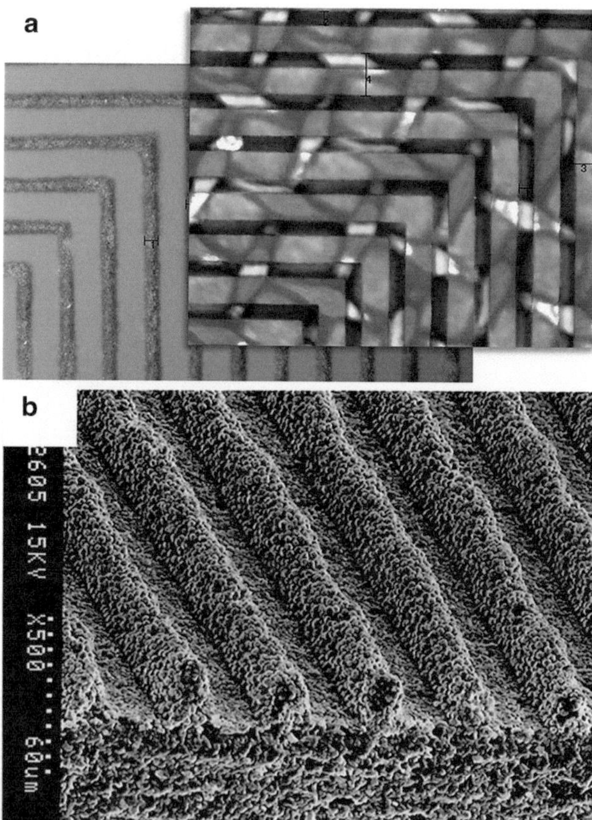

Fig. 2.9 Fine line screen printing (Courtesy of Nakanuma Art Screen, Kyoto, Japan). (**a**) 8 m width Ag nanoparticle ink pattern and screen mask. (**b**) High aspect ratio: 19 μm height, Cu particle ink patterning with L/S = 20/20 μm

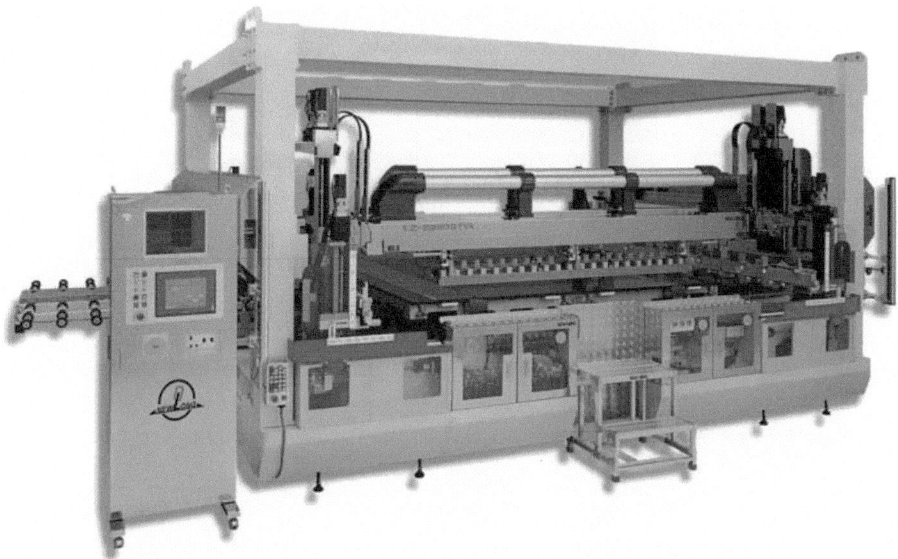

Fig. 2.10   Screen printer for large-scale PDP panel manufacturing (Courtesy of Newlong Machine Works, Tokyo, Japan)

The printing of fine lines of line/space (L/S) below 10 μm/10 μm is possible at the laboratory scale. However, for mass production, current realistic screen printing provides a fineness of 50 μm in L/S production and is expected to reach 30 μm for L/S in the near future. On the other hand, thin printing or coating cannot be achieved in screen printing.

Large-scale screen printing, beyond widths of 2 m as shown in Fig. 2.10, has also been achieved in the industry, especially for plasma display panels, which, unfortunately, are no longer a part of standard TVs.

In screen printing, as schematically shown in Fig. 2.11, printing is performed at a low printing pressure using a screen mesh with a designed pattern of uniform thickness. A flexible metal squeegee or rubber squeegee is used for squeezing paste through the mesh. A polymer mesh, such as polyamide/polyester, or a stainless steel mesh can be used. A mesh pattern is formed by photolithography of an emulsion on the mesh. Instead of a mesh screen with an emulsion pattern, a metal screen can also be used.

Although screen printing is relatively slow, as shown in Fig. 2.1, rotary screen printing, which is used nowadays in large-scale mass production, is very fast, equivalent to other methods of high-speed printing. The resolution of rotary screen printing is, however, limited. Figure 2.12 shows a typical rotary printer with its printing mechanism.

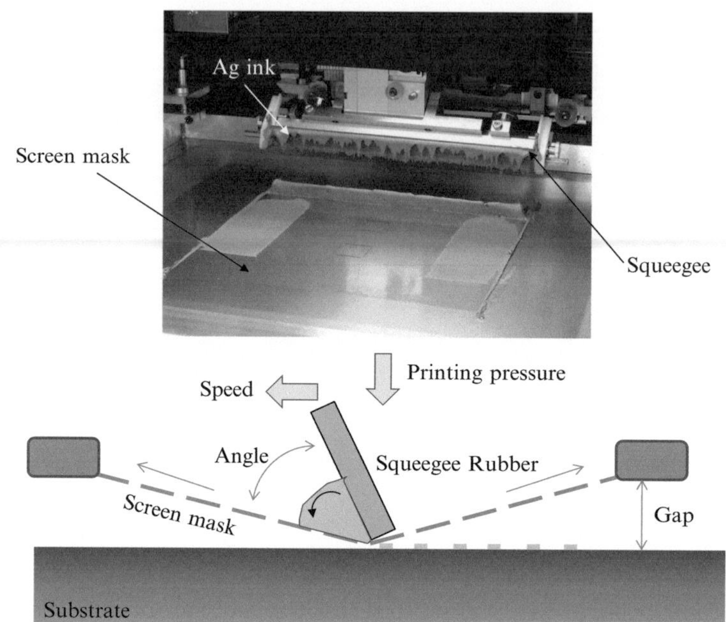

Fig. 2.11 Parameters in screen printing

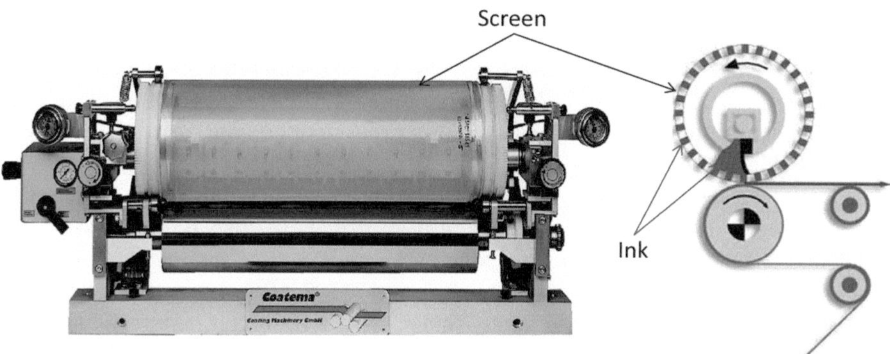

Fig. 2.12 Rotary screen and its mechanism (Courtesy of Coatema Coating Machinery, Dormagen, Germany)

2.3 Inkjet Printing

A piezo drive inkjet has been widely applied to PE technology in a variety of inkjet methods because of its excellent compatibility with functional inks. Inkjet-printed display products have been available on the market. Inkjet technology, which has been around for many years, and its mechanism of droplet ejection are well understood. Figure 2.13 shows a cartoon of ink droplet ejection simulated by a

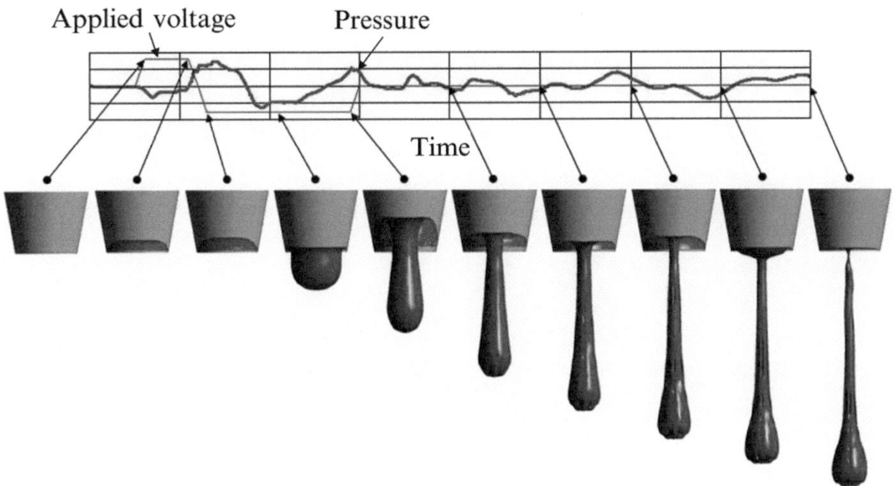

Fig. 2.13 Inkjet droplet simulation by finite-element method (Ansys, Courtesy of Cybernet Systems, Tokyo, Japan)

finite-element method. To form a wiring homogeneous fine line, at each step, the inkjet parameters should be controlled for each nozzle.

The droplet size, shape, speed, and uniformity of an inkjet printer varies from one inkjet head to another, or perhaps from one nozzle to another, even in a single inkjet head. It is necessary to understand the characteristics of an inkjet head and printer algorithm. For instance, a piezo drive waveform, frequency, and amplitude define the initial droplet nature, i.e., shape, size, and speed. Upon droplet ejection, not only the viscosity of the ink and the wettability of the ink on the head material (orifice), but the size and shape of the nozzle tip also affect the amount and shape of the ejected droplet. The shape and direction of droplets during flight also vary greatly depending on ejection conditions. Thus, these ejection parameters should be precisely controlled for each nozzle.

Figure 2.14 shows a series of photographs of Ag nanoparticle ink droplets from the same nozzle where the piezo voltage was changed. In this example, the droplets change their form drastically. At higher voltage, droplets apparently split into two initially, the main droplet and the second satellite, but they coalesce when the second satellite catches up with the main droplet before landing.

The distance to a substrate to be printed from a nozzle tip is usually 1–2 mm, but during flight, air resistance will affect the droplet shape, and in addition, the evaporation of solvent will occur at the same time. When droplets land on a substrate, a droplet wets and spreads on it. Now let us consider the case of droplets of a 2 pl ejection. The diameter of the droplet is approximately 16 μm if it is sphere shaped. When the droplet lands, it will spread as a dot 30–60 μm in diameter depending on the wetting conditions.

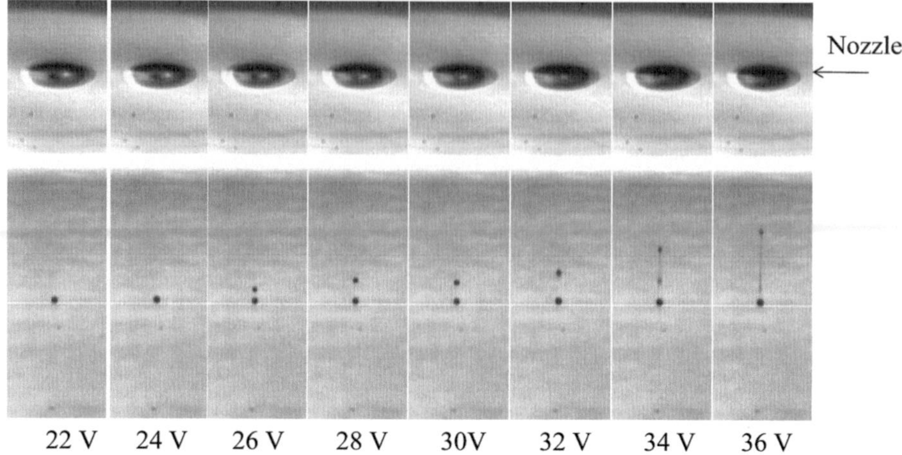

Fig. 2.14 Influence of accelerating voltage on inkjet droplets (frequency: 2 kHz)

Fig. 2.15 Weight of droplet as a function of acceleration voltage

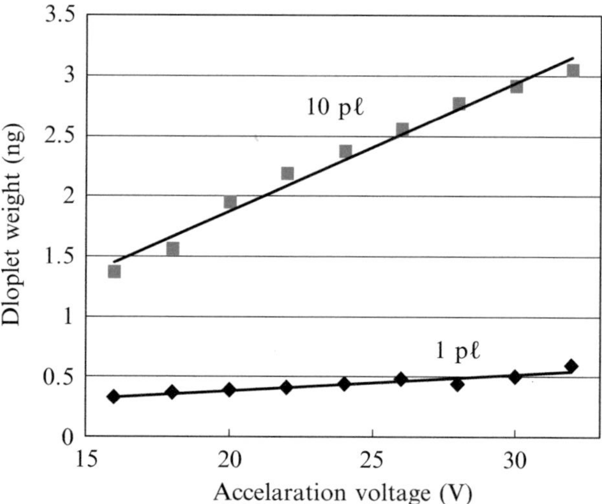

Figure 2.15 shows the variation in droplet weight and velocity that occurs by changing the piezo voltage. Both the speed and weight of the droplets increase linearly with voltage. Thus, conversely, it is possible to reduce the pattern size even using the same nozzle by decreasing the applied voltage.

Controlling the algorithms of inkjet ejection and of stage motion with a substrate is also a key factor in achieving fine patterning. Figure 2.16 shows the applied voltage effect on inkjet patterns. A dot at a piezo voltage of 32 V exhibits an ellipse due to the long tail shown in the photograph, while that at a piezo voltage of 17 V shows a clear circle, as desired.

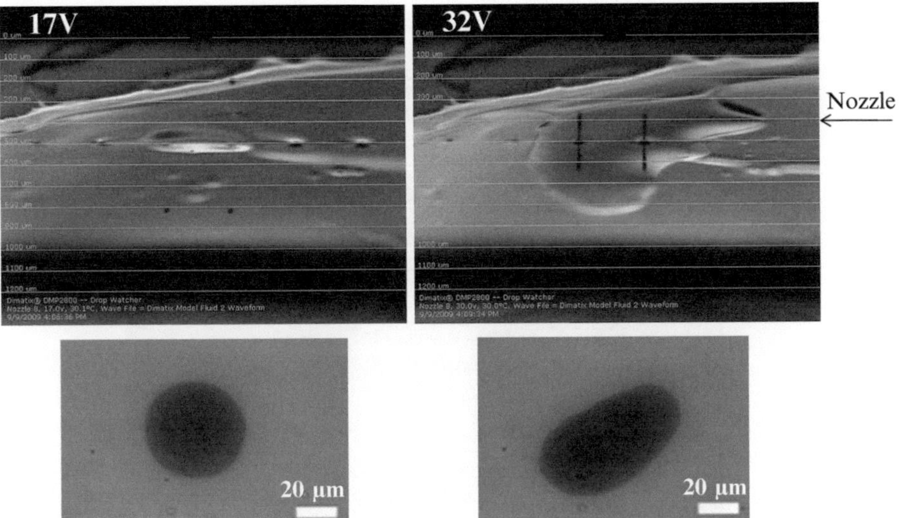

Fig. 2.16 Effect of acceleration voltage on dot shape on substrate

Inkjet printing sometimes unexpectedly forms extra dots that spread out from main patterns, which should be taken into account as the limit of digital imaging technology. A modification must be made to the template for printing images, especially for angled or curved line/edge formation.

When all parameters are suitably controlled, the accuracy of inkjet printing is excellent. Figure 2.17 compares an OLED pixel image before and after adjustment. Because each nozzle has its own deviations, even in a single head, driving each nozzles should be precisely controlled individually. For mass production, an minimum line width/space for typical inkjet printers is 50 µm/50 µm. The accuracy of dots forming on a substrate can be controlled within ±5 µm.

2.4 Fast Printing: Flexo Printing and Offset-Gravure Printing

Flexo printing, which is a very fast relief printing method, has been widely used for flat panel display printing. The mechanism of flexo printing is shown in Figure 2.18, which is suitable for flexible substrates because of the lighter printing pressure involved. The viscosity of flexo printing inks is rather low as compared with those of screen and offset printing; thus, flexo printing has been applied in large-area thin and uniform coating.

Offset-gravure or gravure printing also has an outstanding feature for high-speed mass production. The mechanism of this kind of printing is schematically illustrated in Fig. 2.19. First, ink is placed on a gravure roll of metal and the excess ink is

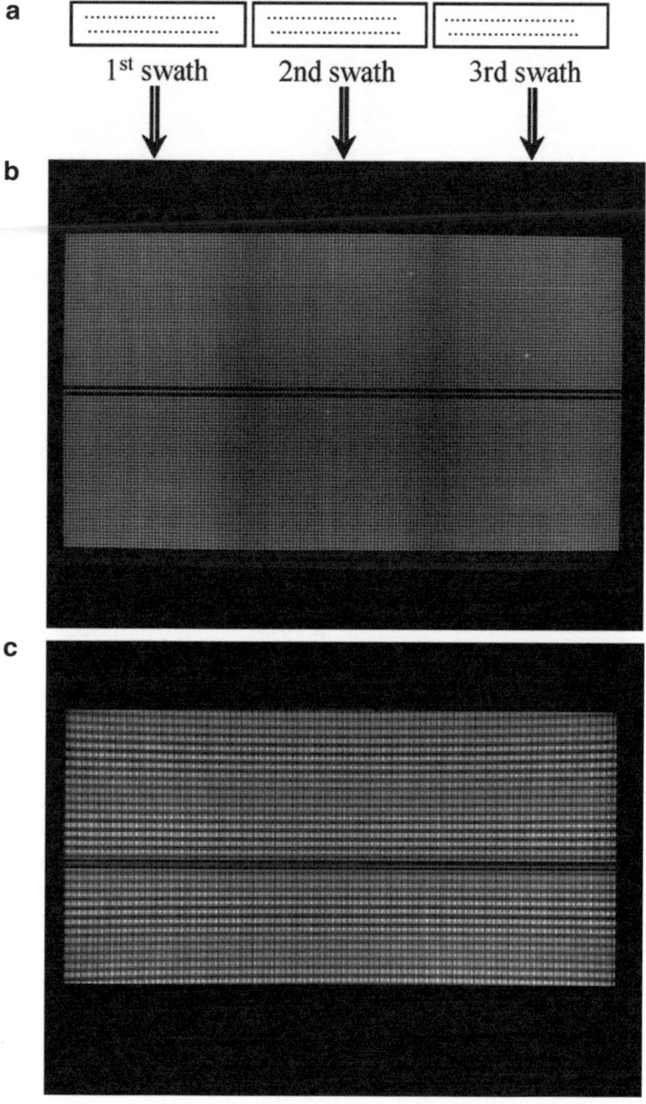

Fig. 2.17 Influence of droplet volume control of each nozzle on OLED pixel images (Courtesy of EPSON, Nagano, Japan). (**a**) Arrangement of heads, (**b**) before adjustment, and (**c**) after adjustment

scraped off with a doctor blade. In the offset process, ink is transferred to a transfer roll, and then ink is finally printed on a substrate under a given pressure. It is possible to accumulate wiring several microns in height depending on its relief depth at very high speeds of up to 1,000 m/min. Because of the softness of the rubber layer on the transfer roll, offset-gravure printing is also very suitable for printing patterns on three-dimensional surfaces with high height steps.

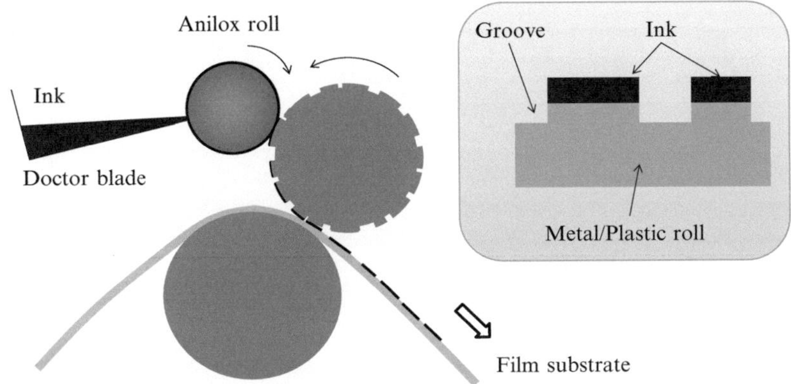

Fig. 2.18 Mechanism of flexographic printing

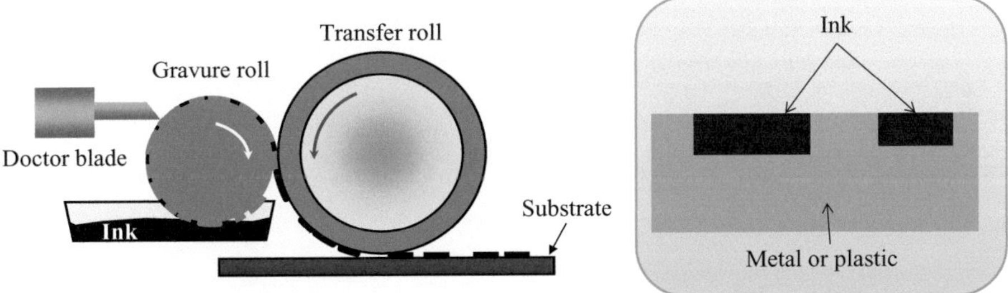

Fig. 2.19 Offset-gravure printing

Figure 2.20 is an example of a gravure plate, withAg nanoink on the plate and Ag printed lines on a substrate. For a preparation of gravure plates, etching is usually performed by photolithographically multiplying copper plating or chromium coating on the roll. In the configuration of the gravure plates, a hard coating like diamondlike carbon (DLC) is frequently applied to surfaces to confer abrasion resistance.

Ideally, in printing, all of the ink on the flexo/gravure plate should be ultimately transferred onto the substrate surface. How this is done is determined by various parameters. Some of the key parameters are listed as follows (Fig. 2.21):

- Materia-l and state of roll/plate: affinity with inks, swelling, hardness
- Ink characteristics: type of solute and its content, viscosity, solvent type/volatility/ amount, absorption by silicone, bubbling
- Depth and pattern/shape of relief
- Wetting and affinity between each roll and ink, surface state of substrate
- Contact pressure of print and transfer roll: push depth (roll deformation), rotational speed of each roll
- Materials and hardness state of plate and doctor blade

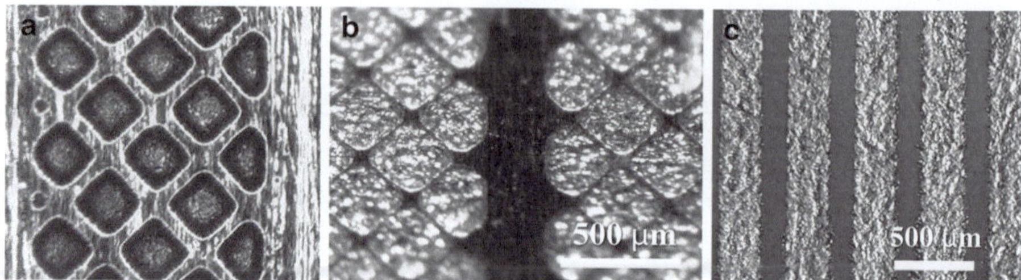

Fig. 2.20 (**a**) Gravure pattern on a roll. (**b**) Ag ink on plate. (**c**) Printed Ag line on paper substrate

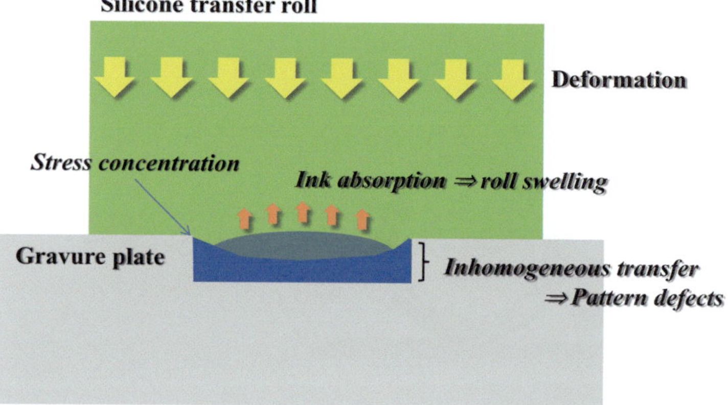

Fig. 2.21 Influencing factors on quality of printed patterns in gravure printing: uniformity of ink, contact of transfer roll under applied pressure, swelling of rubber

For example, in gravure printing, the ink solvent and viscosity, tacking property, applied pressure, and material of the transfer roll significantly affect the ink transfer. Figure 2.21 shows a schematic illustration of the factors to consider in sound printing practices. In particular, an organic solvent is usually used for gravure and offset-gravure printing. The solvent may cause swelling of the silicone transfer roll, such as a permeate silicone blanket, after multiple printings, which will distort the printing quality considerably. Swelling easily occurs when an ink solvent has the same polarity as silicone. The evaporation of the ink solvent also has an influence. Figure 2.22 shows an example of the holding time of ink on a transfer roll [4]. Ink transfer is ideal when enough holding time has passed for solvent evaporation. This shows that the temperature of rolls and plates must be controlled to maintain uniform printing quality.

In mass production, the degradation of a doctor blade that comes from scraping off extra ink causes printing defects and damages plates and transfer rolls. Figure 2.23 shows the doctor blade degradation effects on the formation of many satellite spots. These materials must be selected with great care, especially for mass production.

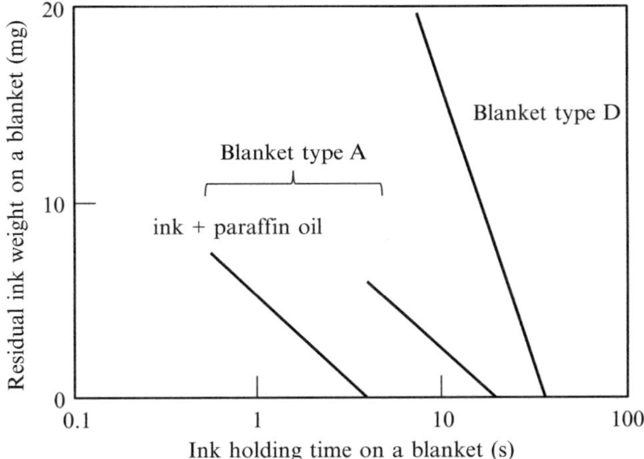

Fig. 2.22 Influence of ink holding time on a transfer roll on residual ink weight on roll blanket [4]. Panels a and d are different types of blanket; the ink contains ceramics particles

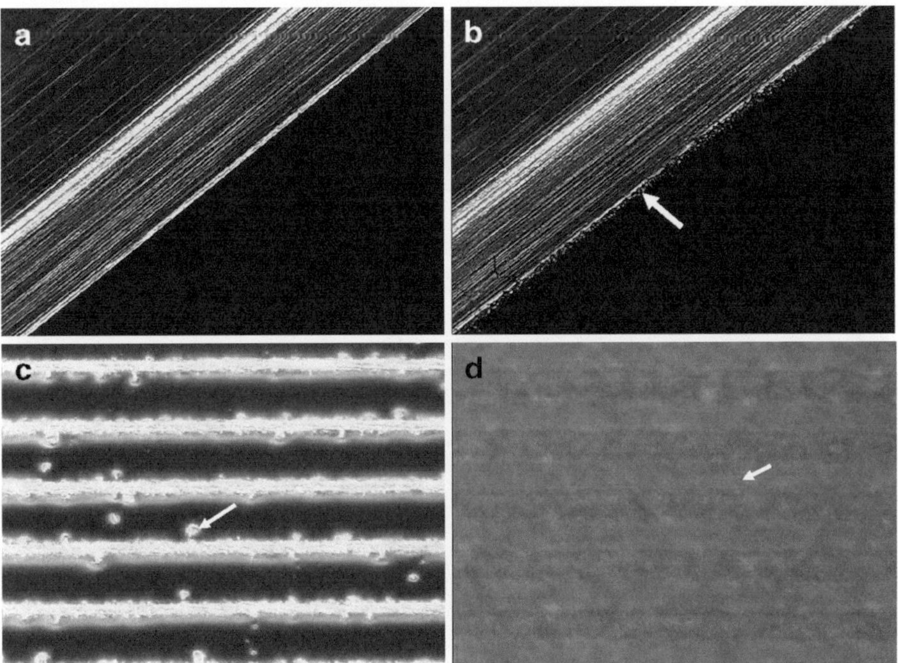

Fig. 2.23 (**a**) Edge of initial doctor blade. (**b**) Edge of degraded doctor blade. (**c**) Ag ink on gravure roll with splash formed by degradation of doctor blade. (**d**) Printed Ag line with satellite spots

2.5   Fine Pattern Printing: Nanoimprint, μCP, and Electrostatic Inkjet

Among fine printing methods, several allow for L/S to be realized even below 1 μm. They include μCP (microcontact printing), nanoimprinting, and electrostatic inkjet printing.

The μCP method is a printing technology that can be applied to fine structure formation down to approximately 100 nm. Kumar et al. first applied μCP in 1993, forming Au wiring with a polymeric flexible stamp [5]. It is known as soft lithography, in imitation of Si photolithography. Figure 2.24 shows a typical process flow of the μCP method. First, the resolution of the patterns is highly dependent on the template [6]. A master template is prepared by photolithography. Polydimethylsiloxane (PDMS, silocone) is commonly used. Thiol is used as a self-assembled monolayer (SAM) that attracts metallic elements such as Au or Ag as in the case or that repels PEDOT or Ni as seen in Fig. 2.24d.

Figure 2.25 shows a fine pattern of Ag nanoparticle ink obtained by National Institute of Advanced Industrial Science and Technology (AIST, Ibaraki, Japan). On flexible plastics such as polycarbonate (PC) or polyethylene naphthalate (PEN), L/S less than 1.0 μm is possible on an area of 15 cm^2. Currently, the μCP method is being used in bio-related technology, such as printing DNA. Unfortunately, even

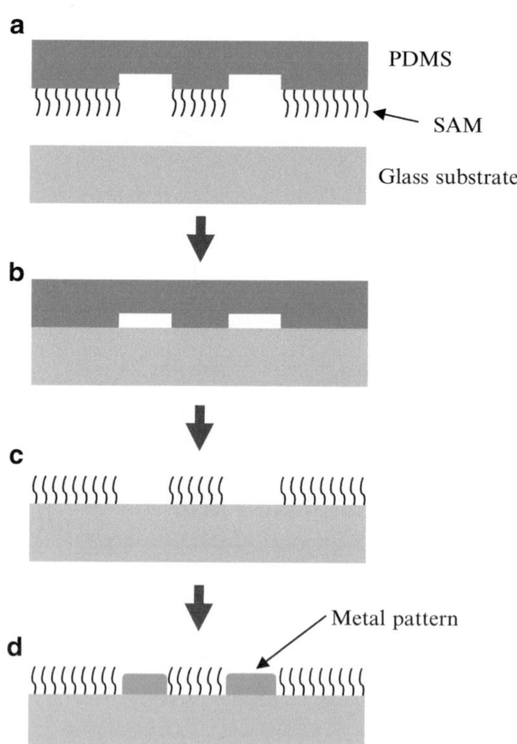

Fig. 2.24 Typical μCP process for forming metal wiring on a glass substrate

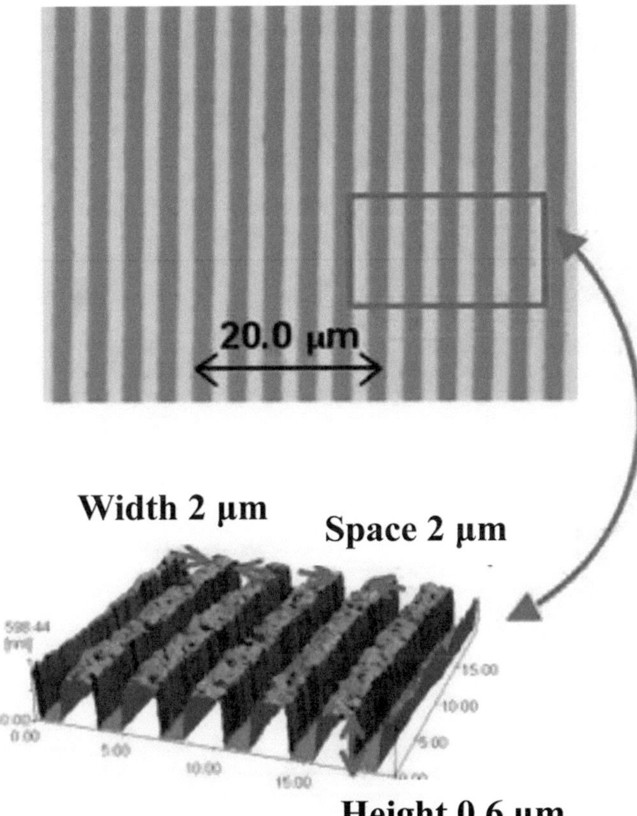

Fig. 2.25 Fine patterns with Ag nanoparticle ink formed by μCP printing (Courtesy of AIST). The *bottom* photo was taken by Atom force microscope (AFM)

though the application expectation is greater in large-area patterning of organic electronic devices, the μCP method cannot meet requirements related to speed and yield in the mass production of PE technology.

Nanoimprinting is also expected to provide submicron pattering, as was first reported by Chou et al. in 1995 [7]. Figure 2.26 show the mechanism of nanoimprinting. This method has been applied to the mass production of hard disk drives (HDDs) and optical films, such as light-guiding or light-scattering plates, but not for the precise patterning of PE technology.

The electrostatic inkjet method is another option for fine patterning, but for a single stroke like dispensing. An ink droplet is driven by kinetic energy in the electrostatic field between nozzle and substrate, as shown in Fig. 2.27. Because the ink is drawn by an electrostatic field in response to voltage applied, it forms a Taylor cone at the tip of the nozzle. A tiny droplet with a high-viscosity ink can be formed without being constrained to the diameter of the nozzle tip. For example, fine lines less than 1 μm in width can be drawn on a substrate by a nozzle with a diameter of 20–100 μm when a high-viscosity ink is used [8].

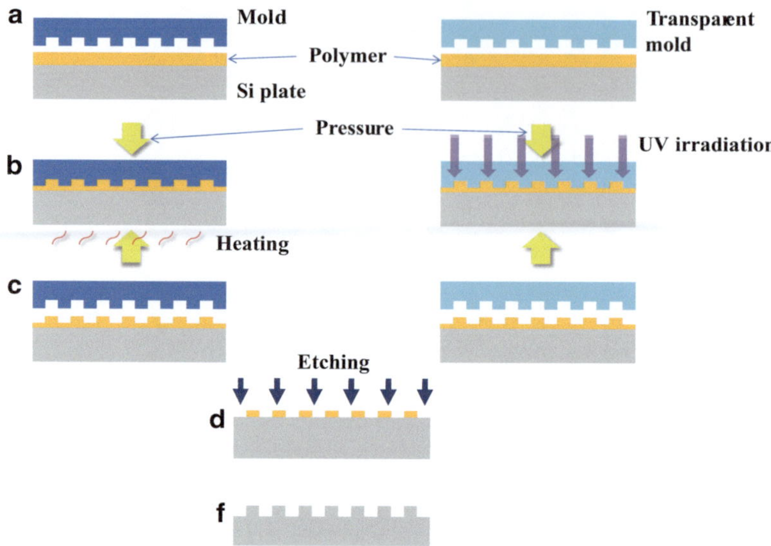

Fig. 2.26 Two different processes of nanoimprint

Fig. 2.27 Electrostatic inkjet printing mechanism. A piezo driver is not necessary

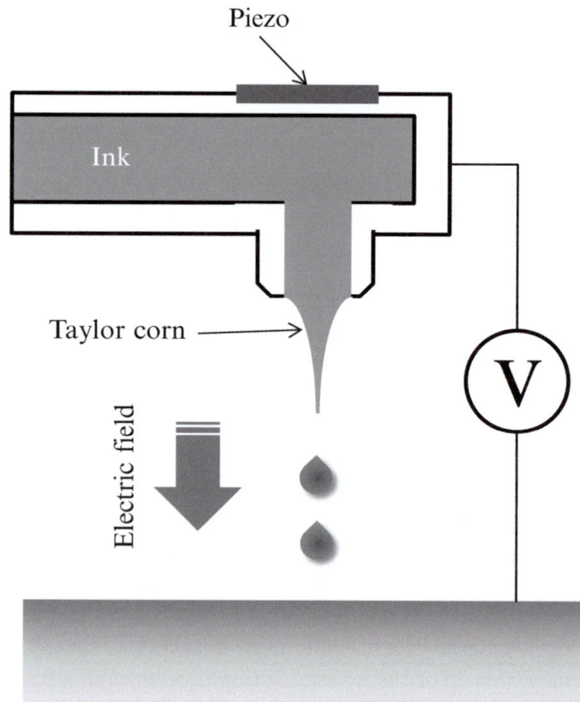

2.6 Laser-Induced Forward Transfer

Laser-induced forward transfer (LIFT) utilizes a metal ablation phenomenon by high-power laser irradiation developed in 1986 [9]. An object film on an optically transparent support is transferred to a substrate by a high-energy focused laser pulse, as schematically shown in Fig. 2.28. The resolution depends on the focus of the laser beam and can be on the order of a few microns. A successful example of the use of this method is shown in Fig. 2.29 [10]. Using the LIFT method, a polymer light-emitting device of a Polymer light emitting diode (PLED)/Al cathode bilayer (poly 2-methoxy-5-2-ethylhexyloxy-1,4-phenylenevinylene, MEH-PPV/Al) was transferred to a silica substrate directly without suffering any damage [10]. The device is uniform and has a very sharp edge. Thus, the LIFT method allows for noncontact, direct-multilayer printing in a solvent-free single step, without

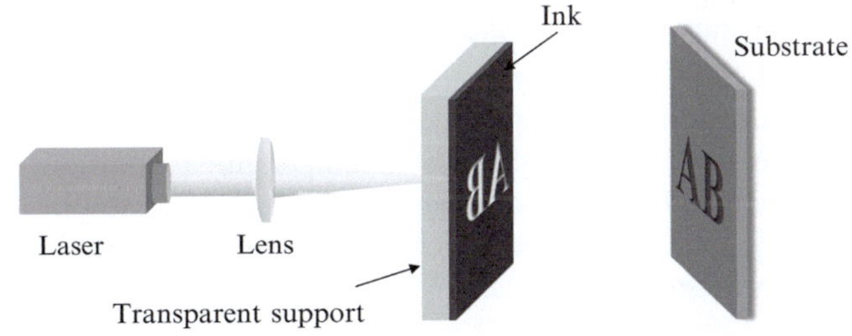

Fig. 2.28 Typical setup for LIFT

Fig. 2.29 View of two pixels through ITO substrate formed by LIFT [10]

requiring any shadowing mask or vacuum installation. The LIFT method is also versatile and can be applied with a variety of donor materials such as metals, organic polymers and monomers, oxide/inorganic compound/Si semiconductors, and even sensitive biomaterials.

2.7 Posttreatment Process

After printing, as shown in Fig. 1.8, printed circuits or devices should be dried or cured before the next step, especially in multiple printing. Drying can be performed with an oven or curing with a UV lamp, both of which are conventional postprinting processes in the printing industry. Functional inks in PE technology have unique requirements in addition to drying. In particular, many metallic and inorganic inks require relatively high temperatures for their densification or crystallization to obtain the desired functional performance. For instance, metallic wiring with Ag and Cu nanoparticle inks requires temperatures exceeding 200 °C to achieve a resistivity of 5×10^{-6} Ω cm. Cu nanoparticle ink further requires an inert atmosphere to prevent severe oxidation. Si nanoparticles or oxide nanoparticles require much higher temperatures, above 300 °C. Such high-temperature treatment will distort the printing process flow and certainly damage most plastic substrates. Instead of high-temperature heating, one should employ certain specific treatments, of which there are several. They are listed as follows:

- Laser curing
- Flash lamp curing
- UV curing
- Plasma treatment
- Microwave curing
- Mechanical forming (cold working)

Direct laser sintering of metal powders is a well-known process involving rapid prototyping technologies [11]. Especially for PE technology, laser curing/sintering is a direct curing or sintering method for ink objects on a substrate. Using the heat energy of a laser, the irradiated pattern increases temperature in a very short time. By adjusting the laser beam size and intensity, one can obtain patterning several microns wide on a heat-sensitive substrate. Figure 2.30 shows an example of laser sintering of a source/drain with Au nanoparticle ink on a Si substrate [12]. As shown in the sequence, first, Au nanoparticle ink was inkjet printed in a wide-track pattern of approximately 100 μm. Then a focused laser was irradiated. The remaining unsintered nanoparticles were washed out, exposing two Au fine lines. A gap between the two lines forms a transistor channel of approximately 4.5 μm.

Flash lamp sintering/curing, which is also called photosintering, utilizes a strong pulsed light irradiation on objects on a heat-sensitive substrate. Figure 2.31 illustrates the mechanism of flash lamp sintering. A strong pulsed light from a controlled Xe lamp through a filter can be absorbed only by an ink object, but not by an optically transparent substrate. Then, only the ink temperature increases without

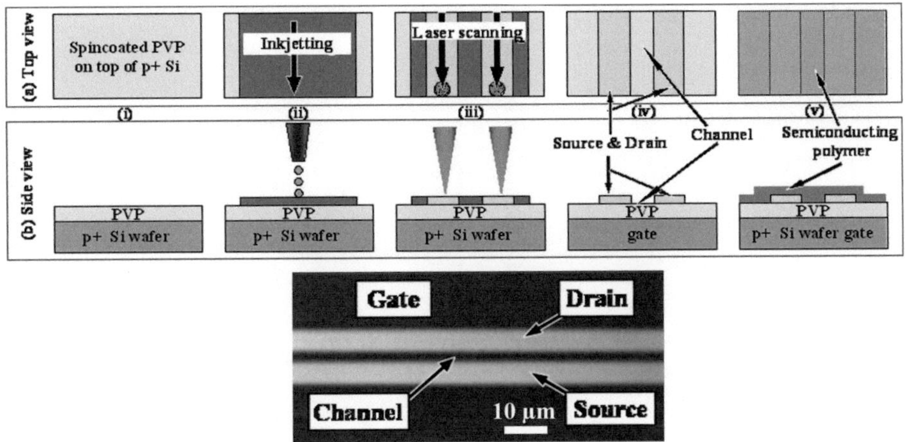

Fig. 2.30 Laser curing of Au nanoparticle ink wiring on Si substrate [12]

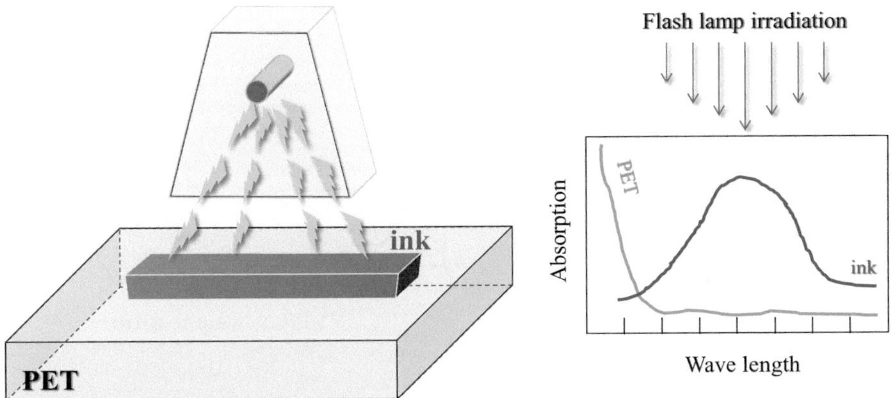

Fig. 2.31 Flash lamp sintering

damaging the substrate. The benefits of this method are the treatment capabilities of a uniform and wide area, its very short run time, and the fact that there are no requirements for a specific atmosphere control such as a vacuum. Even Cu nanoparticles, which are susceptible to oxidation on heating, can be effectively sintered without an inert atmosphere. Figure 2.32 shows the sheet resistance change as a function of Xe flash lamp energy [13]. It is obvious that, as the light energy increases, the sintering of Cu nanoparticles effectively proceeds. Cu nanoparticles are usually covered by a thin oxide layer. In this process, it is likely that the Cu oxide is reduced, absorbing the light energy. In addition to the self-reduction of Cu oxide, a Polyvinylpyrrolidone (PVP) layer covering the Cu nanoparticles greatly influences the sintering, as shown in Fig. 2.33. There is a minimum resistivity in the PVP/Cu

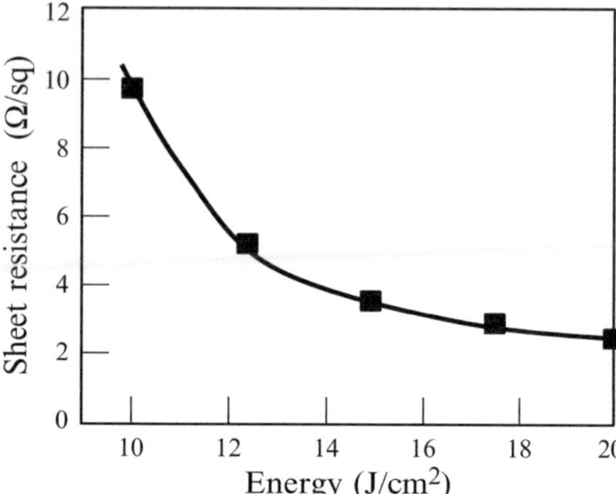

Fig. 2.32 Sheet resistance of flashlight-sintered Cu nanoparticles as a function of irradiation energy on a polyimide substrate [13]

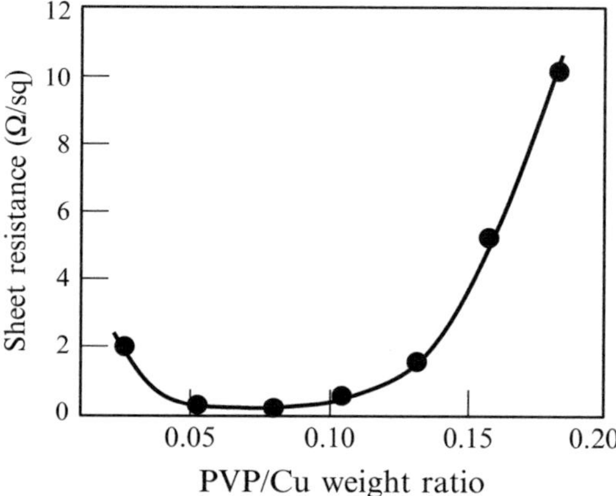

Fig. 2.33 Sheet resistance of flashlight-sintered Cu nanoparticles as a function of PVP/Cu weight ratio [13]

weight ratio range, i.e., from 0.05 to 0.10. The reoxidation of Cu in the low content of PVP has been attributed to this slight increase in sheet resistance. With a higher content of PVP, the resistivity again increases sharply. Flash lamp sintering is also effective in nanowire networks, which will be discussed in the following chapter.

Like flash lamp sintering, UV curing with LED lamps is expected to be effective with PE technology. UV curing has many benefits, such as the absence of infrared in the spectrum, uniform radiation across the exposure width, low cost and long service life, low-voltage operation, instant on/off, and compact size. However, one must keep in mind that the wavelength range in UV curing is in the absorption range for plastic films, which may damage such films.

Plasma treatment is another selective sintering method for metallic inks that uses low-pressure argon plasma. This process shows a clear evolution starting from a sintered top layer into the bulk material, as shown in Fig. 2.34 [14]. Resistivity decreases as the plasma treatment time increases, and the final value is

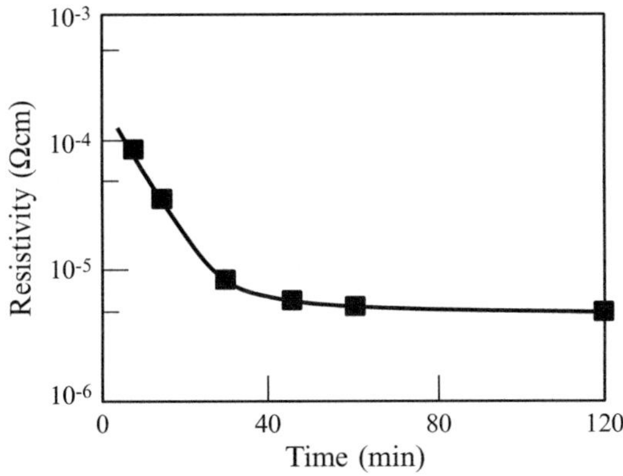

Fig. 2.34 Resistivity change in Ag nanoparticle ink track as a function of plasma treatment time [14]

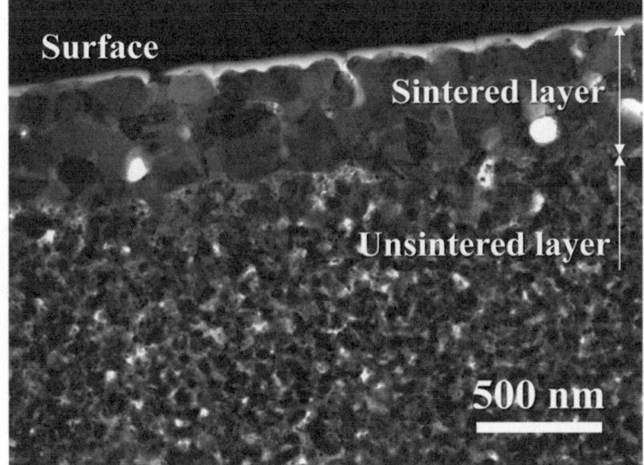

Fig. 2.35 Transmission electron microscope (TEM) of surface of printed Cu nanoparticle ink treated by plasma (Courtesy of Prof. R. Izumi, Kyushu Institute of Technology, Fukuoka, Japan)

approximately three time higher than that of the bulk Ag. Plasma treatment is limited to objects of a certain thickness, which can be correlated to the penetration depth of the plasma into the objects. Figure 2.35 shows a typical example of a cross section sintered by plasma treatment.

The use of microwave radiation is also effective in sintering metallic/inorganic materials. Figure 2.36 shows conductance change as a function of microwave treatment time [15]. The conductance sharply increases after 100 s and almost saturates beyond 100 s. This treatment time shortening is a great advantage of microwave heating. However, metals have a very small penetration depth; the penetration depth of 2.45 GHz microwaves for metal powders of Ag and Cu is 1.3 μm and 1.6 μm, respectively [16]. The conductance or resistivity attained, 3×10^{-5} Ω cm, is approximately 20 times higher than that of bulk Ag.

Mechanical forming is another cost-effective method of PE technology. This will be introduced in the last part of the next chapter.

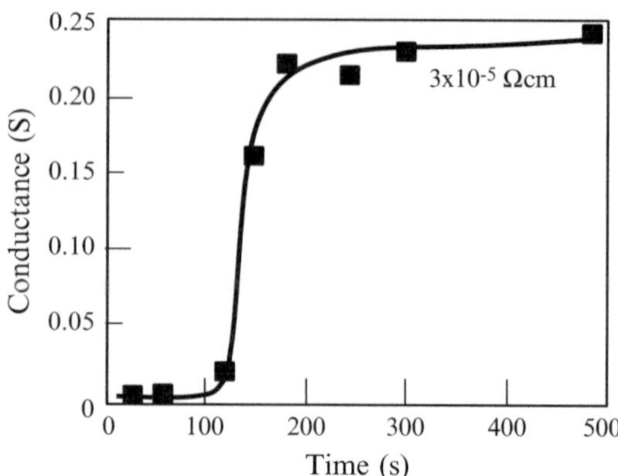

Fig. 2.36 Conductance of printed Ag nanoparticle ink track as a function of microwave treatment time [15]

References

1. Based on OE-a White Paper "Roadmap for Organic and Printed Electronics", 4th edition 2011
2. Sele CW, von Werne T, Friend RH, Sirringhaus H (2005) Lithography-free, self-aligned iInkjet printing with sub-hundred-nanometer resolution. Adv Mater 17(8):997–1001
3. Kim CJ, Nogi M, Suganuma K (2012) Absorption layers of ink vehicles for inkjet-printed lines with low electrical resistance. J Micromech Microeng 2:8447–8451
4. Pudas M, Hagberg J, Leppävuori S (2002) The absorption ink transfer mechanism of gravure offset printing for electronic circuitry. IEEE Trans Electron Packag Manuf 25:335
5. Kumar A, Biebuyck HA, Whitesides GM (1994) Patterning self-assembled monolayers: applications in materials science. Langmuir 10(5):1498–1511
6. Parashkov R, Becker E, Riedl T, Johannes H-H, Kowalsky W (2005) Large area electronics using printing methods. Proc IEEE 93(7):1321–1329
7. Chou SY, Krauss PR, Renstrom PJ (1995) Imprint of sub-25 nm vias and trenches in polymers. Appl Phys Lett 67:3114–3116
8. Ishida Y, Hakiai K, Baba A, Asano T (2005) Electrostatic Inkjet Patterning Using Si Needle Prepared by Anodization J. J Applied Phys 44(7B):5786–5790
9. Bohandy J, Kim BF, Adrian FJ (1986) Metal deposition from a supported metal film using an excimer laser. J Appl Phys 60(4):1538–1539
10. Fardel R, Nagel M, Nüesch F, Lippert T, Wokaun A (2007) Fabrication of organic light-emitting diode pixels by laser-assisted forward transfer. Appl Phys Lett 91:061103
11. Kumar S (2003) Selective laser sintering: A qualitative and objective approach. JOM 55(10): 43–47
12. Kol SH, Pan H, Grigoropoulos CP, Luscombe CK, Fréchet JMJ, Poulikakos D (2007) Air stable high resolution organic transistors by selective laser sintering of ink-jet printed metal nanoparticles. Appl Phys Letters 90(14):141103
13. Hwang H-J, Chung W-H, Kim H-S (2012) In situ monitoring of flash-light sintering of copper nanoparticle ink for printed electronics. Nanotechnology 23:485205
14. Reinhold I, Hendriks CE, Eckardt R, Kranenburg JM, Perelaer J, Baumann RR, Schubert US (2009) Argon plasma sintering of inkjet printed silver tracks on polymer substrates. J Mater Chem 19:3384–3388
15. Perelaer J, de Gans B-J, Schubert US (2006) Ink-jet printing and microwave sintering of conductive silver tracks. Adv Mater 18(16):2101–2104
16. Perelaer J, Smith PJ, Mager D, Soltman D, Volkman SK, Subramanian V, Korvinkdf JG, Schubert US (2010) Printed electronics: the challenges involved in printing devices, interconnects, and contacts based on inorganic materials. J Mater Chem 20:8446–8453

Chapter 3
Conducting Materials for Printed Electronics

3.1 Variety of Conducting Materials

Wiring is one of the essential technologies not only for PE but for all kinds of electronic products. There are several types of electric conducting materials for PE technology. They include metallic materials such as Ag, Cu, and Au, organic molecules such as PEDOT/PSS (poly (3,4-ethylenedioxy-thiophene) poly (styrenesulfonate)), and ceramics such as oxides and carbon nanomaterials. As wiring materials, since electrical conductivity is the most important property, metallic materials are the first choice. Table 3.1 compares the electrical conductivity of these materials. The electrical conductivity of carbon materials such as carbon nanotubes (CNTs) and graphene cannot be directly compared with that of other materials because no bulk material property is available for those nanomaterials. The sheet resistances of CNTs and graphene are discussed in a later chapter as transparent conductive films (TCFs). As for electrical conductivity as shown in the table, it can be concluded that metallic materials, especially Ag and Cu, possess the best values among all the different types of conductive material. Nevertheless, organic materials and ceramics must possess certain specific properties. Details about those factors will be summarized in what follows.

3.2 Metallic Nanoparticles

Many processes have been proposed for the fabrication of nanomaterials in the past 2 decades, not only for metallic materials but also for ceramics and semiconductors. They are categorized into three groups, i.e., a vapor process in which bulk materials are melted and evaporated at elevated temperatures followed by condensation, a reduction of organometallic compounds in solutions with/without the aid of some energy source such as ultrasound, and physical crashing of micron-sized particles

K. Suganuma, *Introduction to Printed Electronics*, SpringerBriefs in Electrical and Computer Engineering 74, DOI 10.1007/978-1-4614-9625-0_3, © Springer Science+Business Media New York 2014

Table 3.1 Electrical conductivity comparison of wiring materials for PE

Materials		Electrical conductivity (Siemens/cm)	Notes
Metals	Ag	6.2×10^5	Bulk properties
	Cu	5.9×10^5	
	Au	4.4×10^5	
	Pt	1.0×10^5	
	Ni	1.4×10^5	
Organic	PEDOT/PSS	1–10^3	
Ceramics	ITO	10^3–10^4	Depends on doping/composition, oxygen defects, and crystallinity
			High-temperature treatment required
	CNT	1×10^5 cm²/V s ($\sim 10^4$ S/cm for CNT fiber [1])	Theoretical values of single-wall (SW) CNT electron mobility (electron mobility of Si is approx. 10^3 cm²/V s). Depends on carrier concentration
	Graphene	2×10^5 cm²/V s	One- and two-dimensional materials

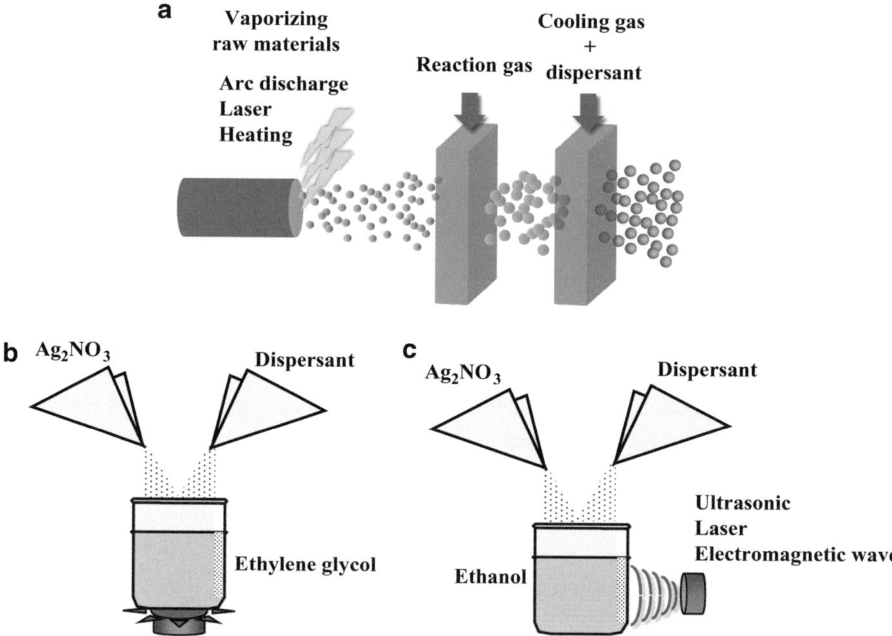

Fig. 3.1 Various nanoparticle synthesis methods. (**a**) Vapor phase process, and two chemical processes. (**b**) Polyol process. (**c**) Reduction with certain energy input

by some mechanical means or by some energy input such as laser abrasion. Typical examples are schematically shown in Fig. 3.1. Of these, the process of chemical reduction in solution has the great advantage of controlling both the morphology and yield of nanomaterials. Reduction agents, which should not be too strong in real-life manufacturing, are added to a solution with raw materials such as metal

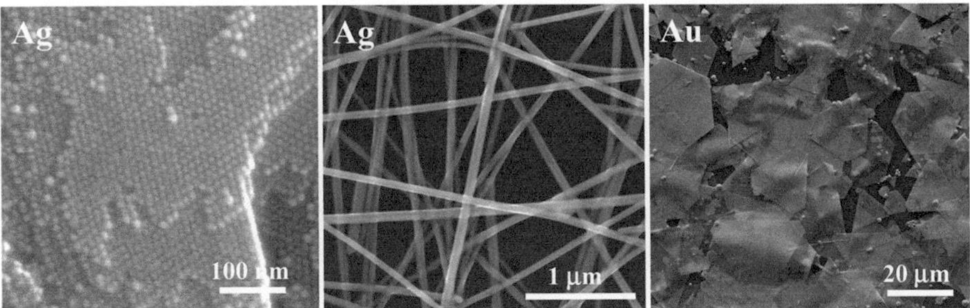

Fig. 3.2 Nanomaterials synthesized by chemical solution process

nitrates or chlorides with certain surfactants. Nanoparticles 1–100 nm in size can be fabricated at very low temperatures, e.g., room temperature up to 200 °C, for a few minutes using solution-based processes, as shown in Fig. 3.2. An appropriate surfactant should be added to preserve nanoparticles at the nanometer scale by preventing their agglomeration. A rod/wire or sheetlike morphology can be achieved by modifying the reaction composition by adding small amounts of suitable agents. These include nanowires such as AgNWs (Fig. 3.2b), which are expected to replace the conventional transparent conductive electrode, Indium tin oxide (ITO), as the first PE mass-produced product.

Now, let us take a look at the attractive features of nanomaterials for PE technology. First of all, because of the nanometer size of these materials are easily formulated into suitable inks that can be used for various printing methods. As mentioned in preceding chapters, low-viscosity dilute inks are required for printing, especially for inkjet, flexo, and offset-gravure printing. In a low-viscosity ink, if particles are too large, the solute particles cannot be dispersed uniformly in a solvent. They easily settle at the bottom of an ink bottle. In contrast, small particles at the nanometer scale can be preserved in a solvent without significant sedimentation due to the Brownian motion. Then, after an ink is made, it can be stored, even on a desk, for months.

The second attractive feature of nanomaterials is the high surface energy state of nanoparticles. In general, the energy state of atoms in a crystalline body is the lowest, whereas atoms on the surface have a higher energy state than those inside the crystal due to dangling bonds on the material surface. As the size of particles decreases, the influence of the surface energy increases, resulting in a high energy state of particles. Buffet et al. measured this effect on the melting temperature of materials using electron diffraction with the aid of simulation [2]. Figure 3.3 shows melting temperature change as a function of particle diameter. Melting temperature evidently decreases as the diameter of particles decreases. In particular, below 20 nm, melting temperature rapidly decreases. The researchers developed two models to explain the melting temperature decrease. One involves the total energy balance between solid-state and liquid-state particles, while the second involves the existence of a surface liquid layer of approximately 0.6 nm. As shown in Fig. 3.3, the diameter needed for a melting temperature of 200 °C is very small, approximately 2 nm. On the other hand, size effects on sintering nanoparticles can be

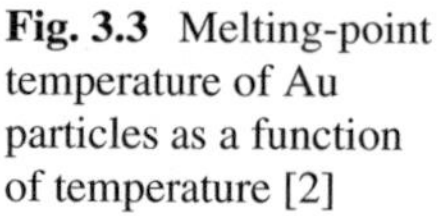

Fig. 3.3 Melting-point temperature of Au particles as a function of temperature [2]

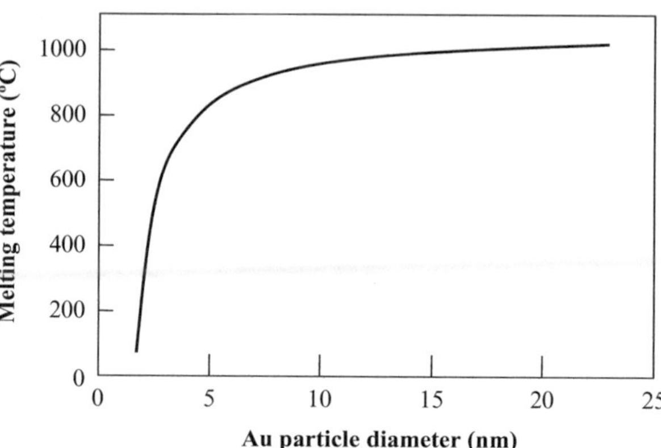

Table 3.2 Typical surfactants for metallic nanoparticles

Nanoparticles	Surfactants
Au	Dodecanethiol, octadecanethiol
	Triphenylphosphine
	Phthalocyanine
	Polyvinylpyrrolidone (PVP)
	Polyethylene glycol (PEG)
Ag	Dodecanamine
	Phthalocyanine
	Polyvinylpyrrolidone (PVP)
	Polyethylene glycol (PEG)
Cu	Polyvinylpyrrolidone (PVP)
	Polyethylene glycol (PEG)

observed even for larger nanoparticles of up to 100 nm. Thus, it is better to say that atomic diffusion becomes very active in nanoparticles near the surface.

Metallic nanoparticles are usually formulated by being stabilized with a thin organic layer or an oxide layer on them. Organic layers can be monomers or polymers suitable for each core metal. Table 3.2 lists some of the typical surfactants used for Au, Ag, and Cu in the literature. In physical processes such as vacuum evaporation and laser abrasion processes, these surfactants are added immediately following the nanoparticle formation sequence. In a solution process, however, these surfactant molecules are added to the reaction solution from the beginning since most solution reactions proceed quickly, resulting in uncontrollable and very fast particle growth. The other choice for surfactants is polymers such as PVP. Polymeric surfactants usually work with most inorganic nanomaterials since they themselves cover nano objects well by forming networks. Nevertheless, polymeric surfactants, since their networks are so strong, are usually not very easily removed at the sintering stage in certain PE production processes by low-temperature treatment. The use of high-temperature curing will have limited application in PE technology.

It is possible to obtain nanoparticles 1 nm to several hundred nanometers in diameter using solution-based processes. Brust et al. first reported a simple

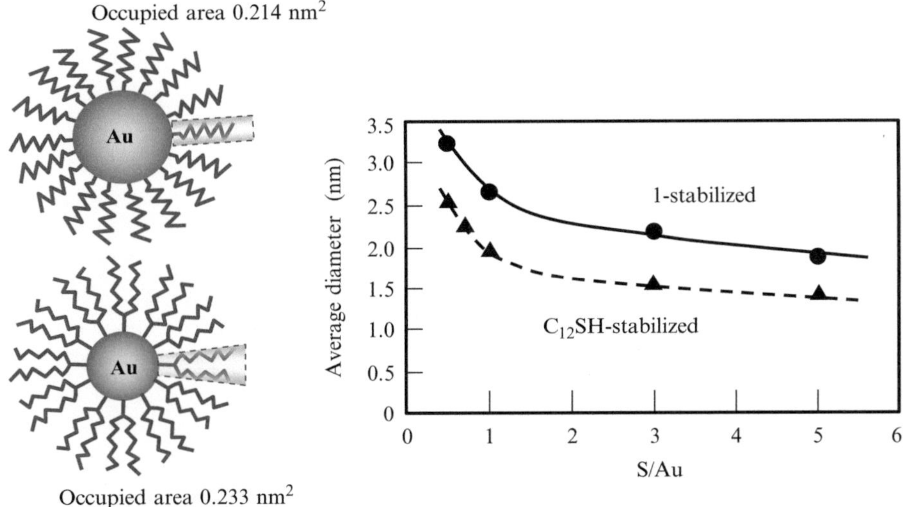

Fig. 3.4 Schematic illustration of stabilizing structure of 1- and C$_{12}$SH-stabilized Au nanoparticles. Effect of stabilizer/Au ratios on average diameters of 1-stabilzied and C$_{12}$SH-stabilized Au nanoparticles [4]

solution-based process for synthesizing 1–3 nm Au particles with a surface coating of thiol [3, 4]. Nowadays, it is well understood that Au can form strong chemical bonds with S (sulfur) in thiol molecules, and this fact has been applied in the stabilization of Au nanoparticles to form a specific atomic arrangement on a surface of a Au cluster [4–6]. Size control of Au nanoparticles can be performed by tuning a molecular cross section of surfactant molecules, as shown in Fig. 3.4 [5]. Larger surface molecules effectively restrict the growth of nanoparticles by a geometric effect. Recently, it was reported that planar molecules such as phthalocyanine can also cover nanoparticles effectively to prevent self-sintering [7].

It is well known that Ag nanoparticles have been used for many decades for their antimicrobial properties [8]. The solution process for manufacturing Ag nanoparticles has been well established [9, 10]. Ag nitrates are decomposed with some reduction agents into Ag ions in a temperature range of 150–200 °C. Amine-based molecules have been often used as a surfactant for Ag nanoparticles, which also plays the role of reduction agent in a solution process. It is easy to synthesize a single nanometer to several hundred nanometers of Ag nanoparticles by controlling the reaction conditions. Nevertheless, the detailed mechanism of atomic bonding between alkylamine and Ag has not yet been clarified. Figure 3.5 shows a schematic image of Ag nanoparticles covered with alkylamine molecules.

The fabrication of Cu nanomaterials is different from the corresponding processes for Au and Ag due to its proneness to oxidation in the atmosphere. A vacuum process can avoid severe oxidation when a suitable capping process is employed at the end of the manufacturing sequences. However, an oxidation effect cannot be avoided in solution synthesis. Due to oxidation proneness, a Cu oxide layer is

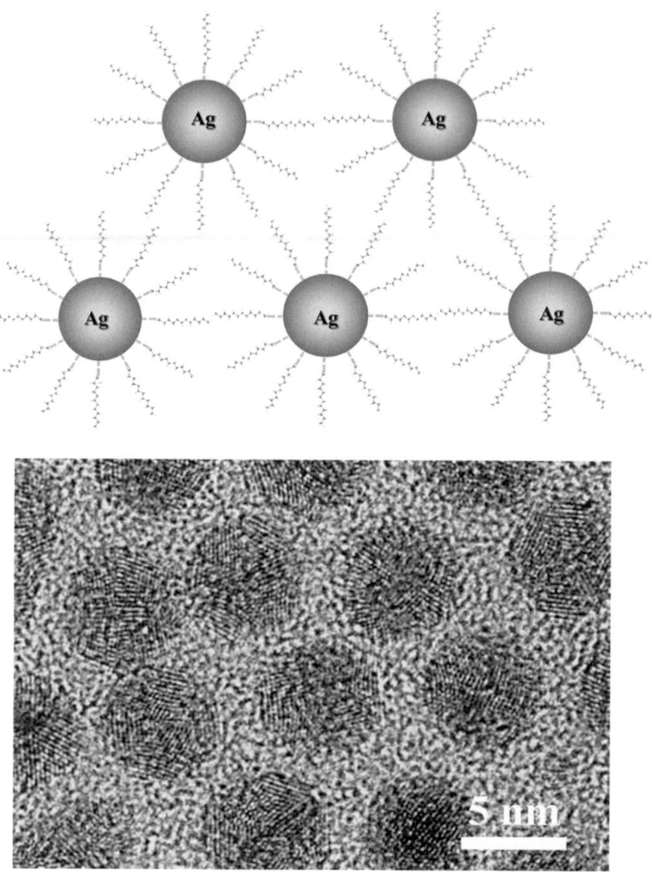

Fig. 3.5 Schematic illustration of Ag nanoparticles and high-resolution TEM micrograph

usually formed on a surface of Cu nanoparticles. Kawasaki et al. reported a successful result of Cu nanoparticle fabrication with a microwave-assisted polyol process without any specific dispersant [11]. Ethylene glycol in a polyol process forms polyethylene glycol (PEG) as a protective layer on Cu nanoparticles.

In the 1970s and 1980s, many metallurgists worked on metallic nanomaterials, especially their nanocrystalline structure and sintering behaviors in a high vacuum. Transmission electron microscopy (TEM) has been one of the common and powerful analysis methods. Even though researchers discovered that nanomaterials could sinter even at room temperature in TEM, no one realized that room-temperature sintering could happen with wiring circuits in ambient atmosphere. A post heat treatment, after printing metallic nanoparticle inks, can provide metallic circuits on substrates. Removing surfactants from printed tracks usually requires heat treatment at a certain high temperature. Figure 3.6 shows a typical example of resistivity change as a function of heat-treatment temperature for Ag nanoparticle ink. Below 200 °C, resistivity is quite high, even for a prolonged heat treatment. At 250 °C,

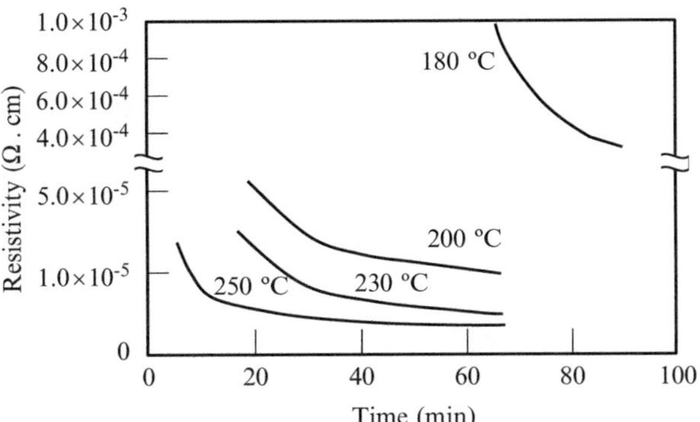

Fig. 3.6 Typical resistivity change of Ag nanoparticle ink with alkylamine surfactant as a function of heating time

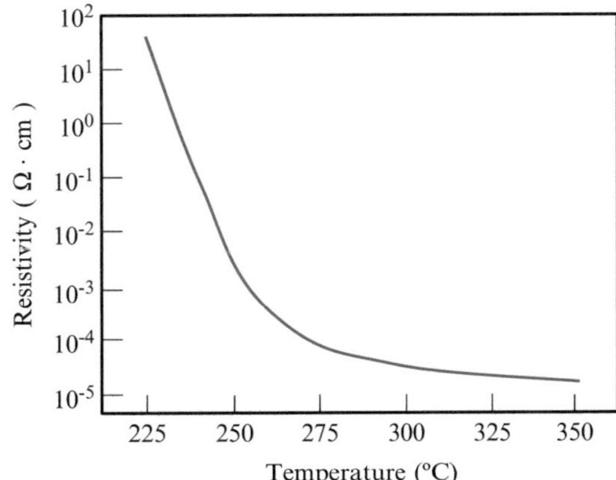

Fig. 3.7 Resistivity change as a function of sintering temperature in vacuum for Cu nanoparticle ink [12]

resistivity decreases rapidly, and the lowest value reaches almost twice that of bulk Ag. This temperature boundary is dependent on the decomposition temperature of the surfactants used for stabilizing the nanoparticles. However, 200 °C is too high a temperature for many of the PE applications mentioned in Chap. 1. Another method must be found to reduce temperature. For Cu nanoparticle inks, in addition to heating treatment for the removal of surfactants, one needs to avoid oxidation during sintering. Usually, a suitable inert atmosphere is employed. Figure 3.7 shows a typical sintering resistivity-temperature relationship curve [12]. A suitable core-shell structure can give the air a stable structure for Cu nanoparticles, where the shell material should be immune to oxidation, e.g., Au, Ag, or Pd. In sintering such core-shell nanoparticle inks, one must consider the fact that the shell layer will be destroyed by a high-temperature treatment. For instance, an Ag protective shell layer on a Cu core nanoparticle is destroyed at around 200 °C, and wiring with the Cu-core/Ag-shell nanoparticle ink still requires an inert atmosphere and high temperature for effective sintering [13].

Wakuda et al. first established the room temperature wiring of an Ag nanoparticle ink [10]. The method is very simple: wash Ag nanoparticles with alcohol to

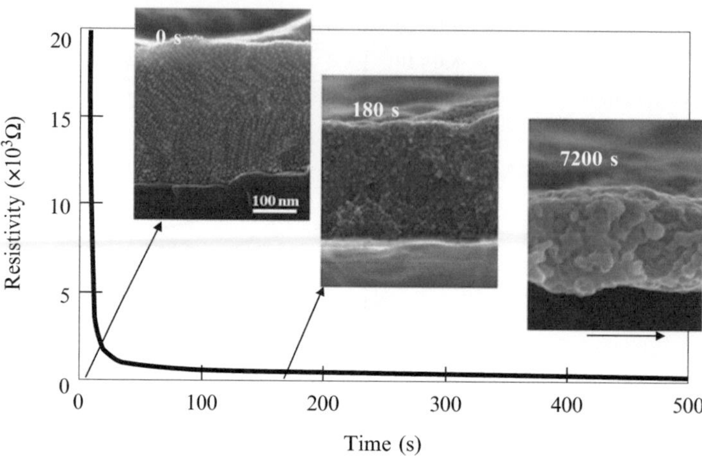

Fig. 3.8 Resistivity and microstructure changes of Ag nanoink wiring after alcohol washing

remove alkylamine dispersant. Figure 3.8 shows the resistivity change in Ag wires printed on a Si substrate as a function of time after alcohol washing treatment at room temperature. Just 3 min is enough to make wire even at room temperature, which demonstrates most clearly the greatest benefit of metallic nanoparticles. Room temperature wiring is also possible using nanoparticles with phthalocyanine as surfactant [7]. In this case, evaporation of solvent makes electrical contacts through the phthalocyanine surface layer, where electrical conduction is not achieved by sintering Ag nanoparticles.

3.3 Metal-Organic Decomposition Ink

Another method for wiring metallic tracks involves solutions with metal-organic precursors where the molecular nature of the metal-organic compound allows a relatively low-temperature formation of the metals. Using this simple concept, the wiring of Au, Ag, and Cu has been developed for printing applications [14–18]. In all cases, printed patterns were converted into continuous, conducting metallic tracks by a postdeposition heat treatment.

Ag-based inks have been frequently used in the form as metal-organic decomposition (MOD) inks. The ink itself does not contain nanoparticles and is transparent, as shown in Fig. 3.9. Among various potential MOD inks (although a limited number of papers has been published on the topic), β-ketocarboxylate Ag ink has the lowest decomposition temperature, approximately 100 °C, originating from its molecular structure, which is shown in the inset of Fig. 3.9. Figure 3.10 shows the resistivity change in β-ketocarboxylate ink as a function of curing time on a flexible plastic substrate [18]. At 120 °C, the resistivity reaches the order of 10^{-6} Ω cm for 60 min.

One of the great advantages of MOD inks is the surface smoothness of the cured pattern, which becomes a mirror, as shown in Fig. 3.11. For instance, the surface

Fig. 3.9 β-ketocarboxylate
and its 100 °C curable
ink [18]

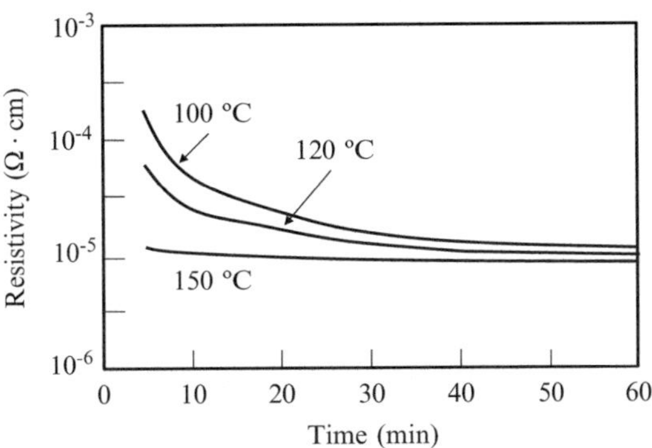

Fig. 3.10 Resistivity change
as a function of sintering time
in air for Ag carboxylate
ink [18]

roughness of a β-ketocarboxylate MOD ink cured at 120 °C on a glass substrate becomes approximately 29 nm of Ra roughness, while those of Ag nanoparticle ink patterns, Ag-epoxy conductive adhesive patterns, and commercial-based Al foils are 50.2 nm, 173 nm, and 28.0 nm, respectively.

Cu-based MOD inks are also attractive for their low cost compared to Ag or Au. Nevertheless, it is very hard to cure a Cu-based MOD ink without exposing it to the

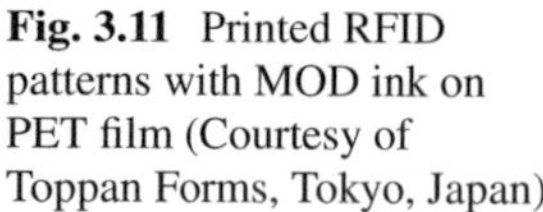

Fig. 3.11 Printed RFID patterns with MOD ink on PET film (Courtesy of Toppan Forms, Tokyo, Japan)

effects of oxidation in an air atmosphere. Without a strong reduction agent, an inert atmosphere and higher temperature are required. Processes in which control of the atmosphere is potentially not as essential are the photosintering methods mentioned in the previous chapter. Araki et al. were able to sinter Cu-MOD inks in an air atmosphere by photosintering [19]. A resistivity on the order of 10^{-5} Ω cm was achieved without a heating furnace.

3.4   Nanowires

Metallic nanowires possess unique properties in the application of electronic wiring. Due to their one-dimensional high-aspect-ratio morphology, electrical connection becomes much smoother than in nanoparticle wiring as well as heat conductivity. TCFs with a random network of nanowires are the typical application utilizing this feature. In addition, nanowire antennas possess a superior high-frequency response.

Nanowires are also called nanorods, nanofibers, and whiskers, and they all have a fibrous morphology. As shown in Table 3.3, typical diameters of nanowires are in the range of 10 nm to several hundred nanometers, where the maximum diameter should be less than 100 nm based on the definition of *nano*. Length varies from a few hundred nanometers up to 50 μm. Nanofibers are almost identical to nanowires, and nanorods are short versions of nanowires. Whiskers are slightly different from the others. They are relatively thick, up to 10–20 μm. Most of them are single crystals, except for metallic nanowires and nanorods. Many different kinds of whiskers have been reported.

Among the various nanowires, the most useful for PE technology are Ag and Cu nanowires. Although several synthesis methods have been proposed in the

Table 3.3 Properties of conductive fibrous nanomaterials

Material	Thickness (nm)	Length (µm)	Electrical conductivity (Ω cm)
Ag nanowire	20–150	1–50	1.6×10^{-6}
Cu nanowire	5–120	10–50	1.7–40×10^{-6}
Ni nanowire	10–200	1–5,000	30–50×10^{-6}
C fiber	100–150	50–100	5×10^{-5}
CNT	0.4–50	1–1,000,000	–
Sn whisker	1,000–20,000	5–10,000	11.0×10^{-6}
SiC whisker	100–1,000	10–100	0.01–0.03

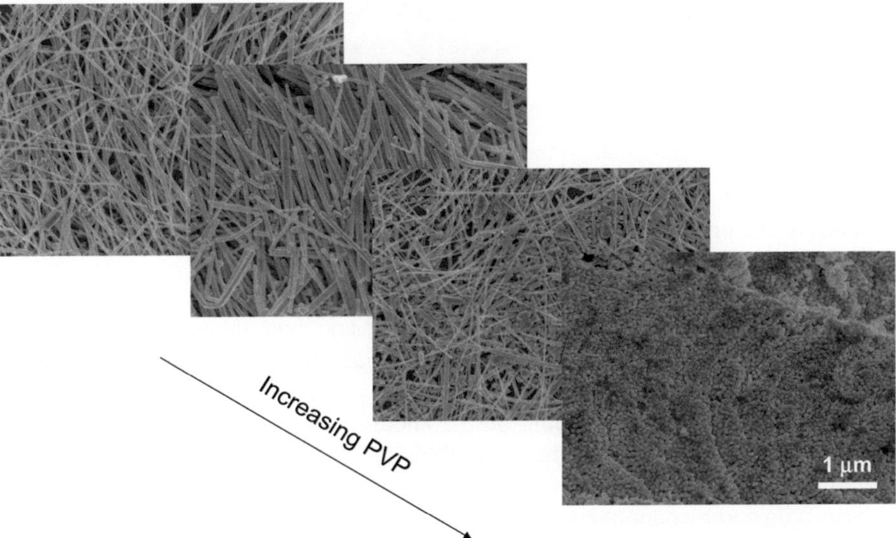

Fig. 3.12 Mass-produced AgNWs fabricated by polyol process

literature, a common synthesis method is well established for nanowires, i.e., a typical polyol process with the aid of a small amount of catalyst. For instance, in the synthesis of Ag nanowires (AgNWs), an ethylene glycol solution of $AgNO_3$ and PVP with a suitable amount of NaCl is prepared at room temperature [20, 21]. The mixture is then heated at 150–200 °C. Instead of ethylene glycol, ethanol or water can be used [22]. By changing the synthetic parameters, composition, temperature, and time, one can control the thickness, length, and yield. Figure 3.12 shows a typical example. After optimization, the yield of AgNWs exceeds 90 % of the raw material. Cu nanowires (CuNWs) are obtained either by reducing $Cu(NO_3)_2$ with hydrazine in an aqueous solution containing NaOH and ethylenediamine (EDA) at 80 °C for 60 min [23] or by reducing $CuCl_2$ with glucose hexadecylamine in an autoclave at 120 °C [24]. All metallic nanowires fabricated by a polyol or aqueous process have fivehold twining along the wire axis. Figure 3.13 shows a typical cross-section image. Such a unique growth mechanism has been primarily explained

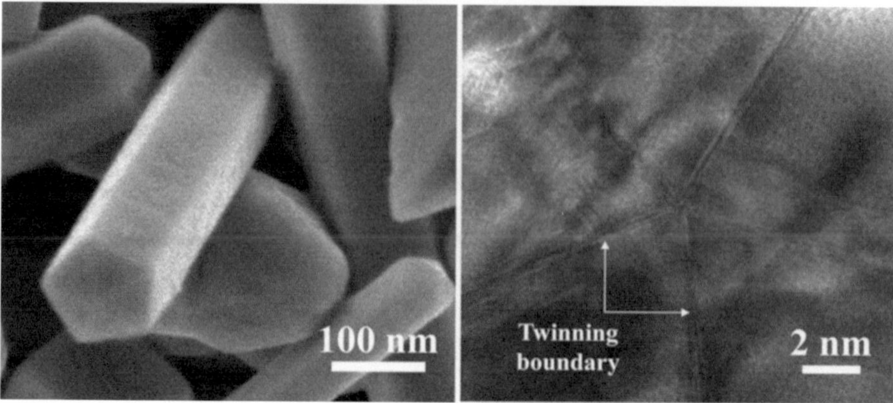

Fig. 3.13 Morphology of AgNWs (SEM)

in terms of the presence of surfactants or capping agents that regulate the growth of a crystal in a particular direction [25].

The template methods are the other nanowire fabrication methods utilizing a solution process [26]. Porous alumina created by anodization on Al in acidic electrolytes serves as a template for nanofibers. The nanofibers are cylindrical and uniformly sized according to the uniformity of pores whose diameter ranges from a few nanometers to 200 nm, depending on the anodization conditions. One could also control the pore morphology, whose length ranges from less than 20 nm to approximately 1 μm. Metal ions fill in the pores after anodization by applying an AC/DC current followed by nanofibers growth by deposition.

3.5 Other Conductive Materials

Organic materials do indeed possess low electrical conductivity as compared to metallic materials (Table 3.1). And though organic materials cannot replace metals, they are attractive due to their ability of ink formulation and of flexibility. For instance, they are soft and flexible, lightweight, inexpensive, very compatible with organic and aqueous solvents, and possess many other features, though some of them are still potential features. Thus, organic materials have great potential as wiring or TCF materials for PE technology. As organic electronic materials, they can be divided into large two groups, i.e., monomers and polymers. Organic electronic polymers are usually considered conductive elements. Monomers, as well as some polymers, are used in semiconducting devices, where a short distant electron/hole mobility is required. These organic semiconductors will be discussed in the next chapter.

In general, since electrons are firmly fixed to the atoms in polymers, electrons cannot flow along the networks of polymers. However, when a polymer has conjugated

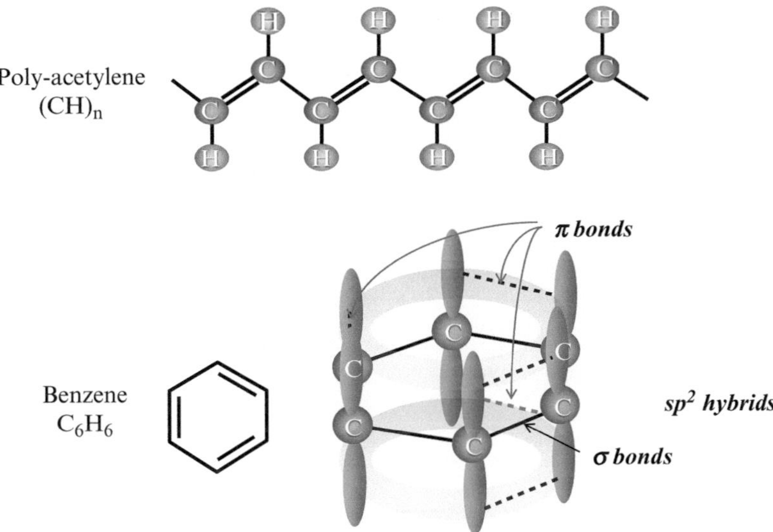

Fig. 3.14 Conductive conjugated double bonds of polyacetylene and benzene

double bonds (continuous double bonds have a π bond network), like polyacetylene or benzene, electrons can move along the π bond network. Figure 3.14 shows the chemical structures of typical conjugated double bonds. This characteristic was discovered over 40 years ago for polyacetylene [27]. In the case of nonconjugated polymers, the carrier is localized to the conjugated units, and the carrier becomes mobile by a hopping mechanism connecting the localized conjugated molecules. Like a Si technology, the doping of various elements increases carrier mobility such as for polyacene, polyaniline, polyparaphenylene, polythiophene, polypyrrole, and various polymers. Some of them have been successfully applied in lithium ion batteries, aluminum electrolytic capacitors, tantalum electrolytic capacitors, and antistatic films.

Among polymer-based conductive materials, superior electrical conductivity characteristics were found for thiophene [PEDOT/PSS, poly(3,4-ethylenedioxythiophene) poly(styrenesulfonate)]. Its molecular structure is shown in Fig. 3.15. Although PEDOT is insoluble intrinsically, it becomes soluble in the presence of polystyrene sulfonic acid (PSS) as a colloidal dispersion in an aqueous solution. Further, the addition of an aqueous solution having the same charge system is also effective as a PEDOT/PSS mixture of polyvinylpyrrolidone (PVP). Its conductivity characteristics have been improved, as shown in Fig. 3.16 [28]. Various types of doping have also been proposed both for conductivity improvement and for increasing stability. Today, a conductivity of up to 1,000 S/cm, which is almost equivalent to ITO, can be achieved. PEDOT/PSS can be characterized as a very stable conductive polymer with respect to temperature and humidity. Nevertheless, environmental stability is one of the issues to consider in typical polymer materials.

One of the goals for PEDOT/PSS development is to take market share away from the TCFs. In other words, to be a standard TCF material, current ITO, is the driving

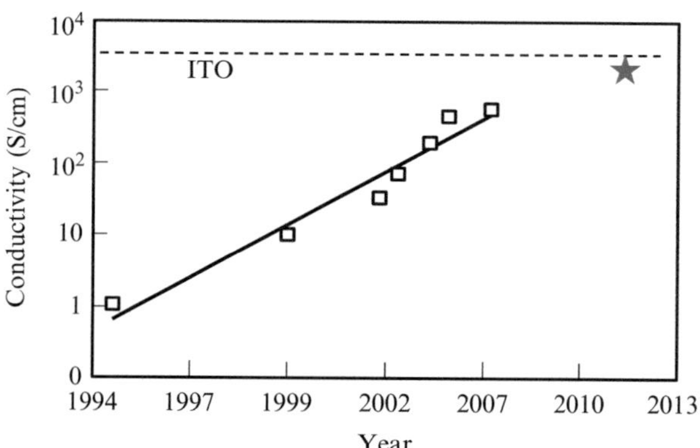

Fig. 3.15 Molecular structure of PEDOT/PSS

Fig. 3.16 Conductivity change of PEDOT/PSS (from a brochure of H.C. Starck). The star was added by the author

force of the research and development. There are great benefits of PEDOT/PSS, i.e., low cost, low resource barrier, flexibility, low process temperature, and compatibility as aqueous solvents. As compared with other competing materials, the drawbacks are their reliability, in terms of light irradiation and exposure to heat and humidity, and their low conductivity. Anyone working with this material should keep these weak points in mind.

3.6 Other Conductive Nanomaterials and Applications to Transparent Conductive Films

Oxide ceramics and carbon nanomaterials represent additional candidates as conductive materials. As conductive oxides, most practical oxides are prepared by adding a dopant to distort either the cation or anion lattice. Typical doped materials are Sn oxide doped with Sb (ATO) or florin (FTO), In oxide with Sn (ITO) or florin, and Zn oxide doped with Al (AZO) or Ga (GZO). Of these, ITO is still the best choice

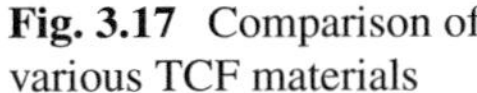

Fig. 3.17 Comparison of various TCF materials

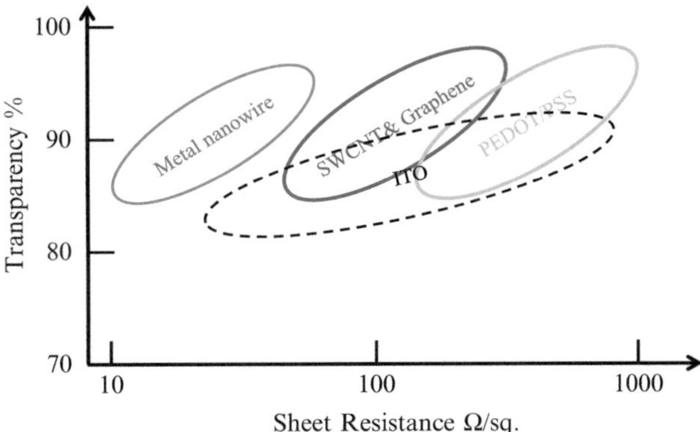

as a conductive electrode. As for carbon materials, nanotubes and graphene are two of the most attractive conductive nanomaterials, while larger carbon materials, such as carbon fibers or particles, are already being used as electrodes or resistors in existing electronic devices. Using these nanomaterials, considerable research efforts have devoted to transparent wiring technology.

The development of organic and inorganic alternatives to a scarce and expensive material, ITO, that is currently being used as a TCF in various types of displays, in photovoltaic cells of amorphous Si/CIGS/organic thin film/DSSC, and in OLED lighting on glass, metal, and plastic substrates. A material is defined as a TCF with a light transmittance that is more than 70 % in the visible spectrum and an electrical sheet resistance that is less than 10^3 Ω/sq., while the requirement for resistivity/transparency varies from one to another depending on applications. ITO's used as an electrode is well established in the majority of current applications. However, its availability forms a barrier to market expansion, particularly for devices with better conductivity, light weight, and greater flexibility on plastic substrates. The new device field requires the development of inexpensive novel conductive materials, more widely available elements, and processes. Novel printing procedures must also be developed to enable direct writing of multi- and single-patterned nanolayers, removing the waste associated with etch patterning. Figure 3.17 shows a comparison of four types of TCF alternatives as a function of sheet resistance and optical transparency.

Doped oxides have been most intensively studied for more than 3 decades using a solution process and sol-gel methods, and in fact, they are industrially applicable coating types. A sol-gel process is a chemical method for making inorganic glasses from a solution of metal alkoxide compounds. The process uses various alkoxide precursors, in the form of a solution or dispersion, that undergo a series of chemical reactions to form first a liquid sol and subsequently a solid gel. Sol-gel methods have been intensively researched for their photovoltaic applications. A variety of precursors have been used for ITO formation. A typical sol-gel coating process is shown in Fig. 3.18 [29]. First, In and Sn source compounds are mixed in ethylene glycol at low temperature to obtain the gel, coating or printing them with solvent on a substrate, and then fired at high temperature. Figure 3.19 shows resistivity as a function of firing temperature. To obtain a resistivity level below 10^{-2} Ω cm, a

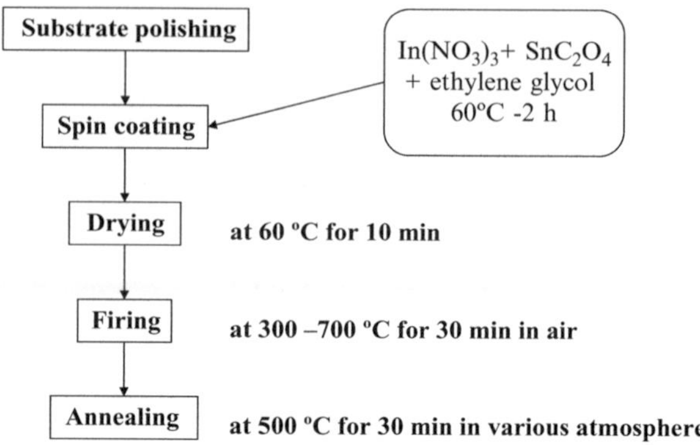

Fig. 3.18 Typical sol-gel coating procedure [31]

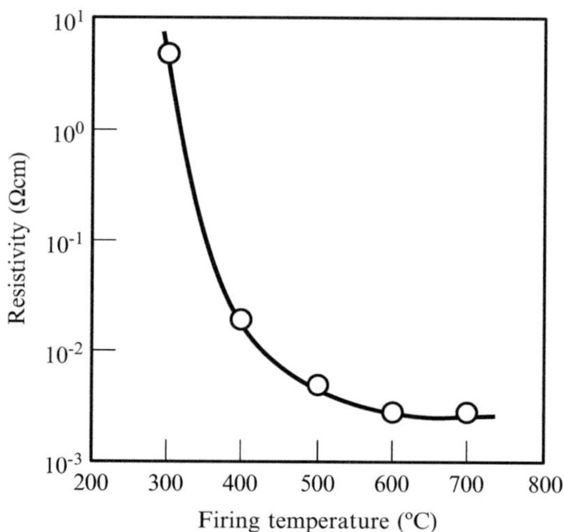

Fig. 3.19 Firing temperature dependence of resistivity of ITO coating [29]. Sn/In = 8 mol % in coatings

temperature above 400 °C is usually required. A firing temperature greater than 200 or 300 °C is not desirable for most PE products. Recently, the great potential for decreasing the temperature in sol-gel methods was reported for ZnO-based oxide semiconductors [30, 31], which will be discussed in the following chapter. This exception is not a crystalline material but amorphous oxides, which were discovered by Hosono over a decade ago [32].

Another method is nanoparticle solution inks [34, 35]. Figure 3.20 shows typical ITO nanoparticles [33]. The drawback of using a nanoparticle ink lies in the difficulty of obtaining dense oxide films in a low-temperature sintering process, even at 400 °C, compared to sol-gel processes. Materials of commercial interest include ITO, ATO, and FTO. The sheet resistance reached almost 100 Ω cm, while heat treatment temperatures can reach as high as 500 °C. In addition, heat treatment requires a controlled atmosphere to adjust for oxygen lattice defects, which provides an electric property to oxides. Figure 3.21 shows an example of resistivity change as a function of sintering temperature [34].

Random networks of CNTs have been an attractive choice for TCF applications due to their excellent stability. CNTs have been proven to be a unique material for

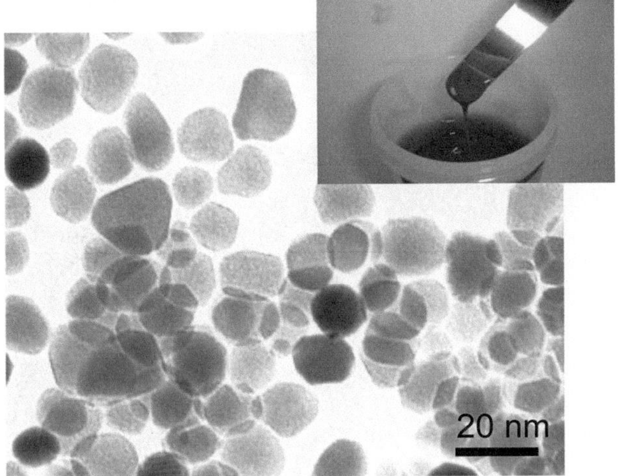

Fig. 3.20 TEM of ITO nanoparticles (Courtesy of Otsuka Chemical, Osaka, Japan). ITO nanoparticles were obtained by a thermal decomposition of metal alkoxide [29]

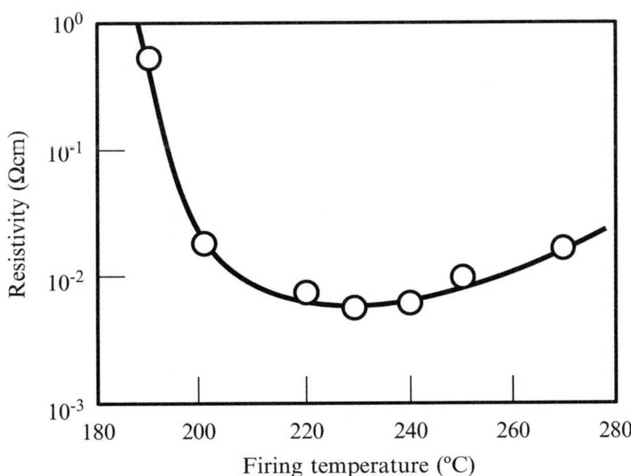

Fig. 3.21 Sintering temperature dependence of resistivity of ITO nanoparticle film [32]. First sintering was carried out at 190–270 °C for 60 min in vacuum followed by annealing at 230 °C for 60 min in Ar

TCFs. Because CNTs have two different conductivity types, metallic conductive and semiconductive, the separation of metallic CNTs from semiconductive-type CNTs is an essential requirement. Although many trials have been reported on the separation of CNTs, none of them has succeeded in the mass production of metallic CNTs yet. Single-walled carbon nanotubes (SWCNTs) are currently the best choice for conductors, even though their high cost will limit applications. CNT networks and metallic nanowires both have the drawbacks of an intrinsic percolation limit and electric contact at two CNTs overlap [35]. A highly transparent and conductive film requires a uniform dispersion of long, thin CNTs with smooth surfaces. The creation of a uniform dispersion of CNTs is the first difficulty. Several strategies can be used to disassemble CNTs in a solution. For instance, mechanical methods, such as high-energy stirring with or without ultrasonic agitation, are commonly used. Surface modification and functionalization of CNTs with a suitable surfactant is also effective. The electrical conductivity of a SWCNT network is limited by highly resistive junctions between SWCNT bundles. Figure 3.22 shows the effect of

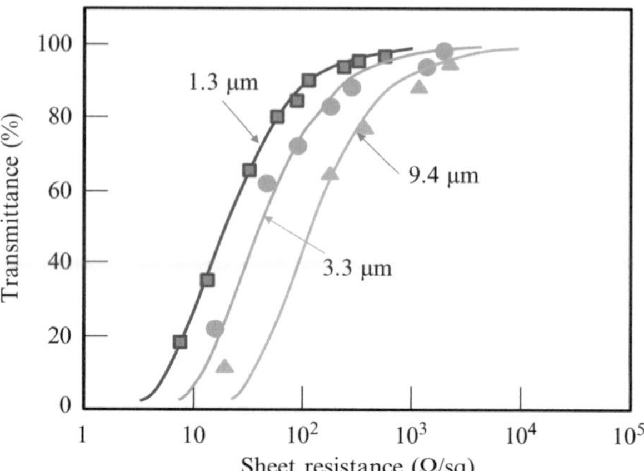

Fig. 3.22 Effect of SWCNT bundle size on sheet resistance–transparency properties [36]. Three bundle sizes of SWCNT samples indicated were used for the preparation. The TCFs were treated with HNO_3 for doping

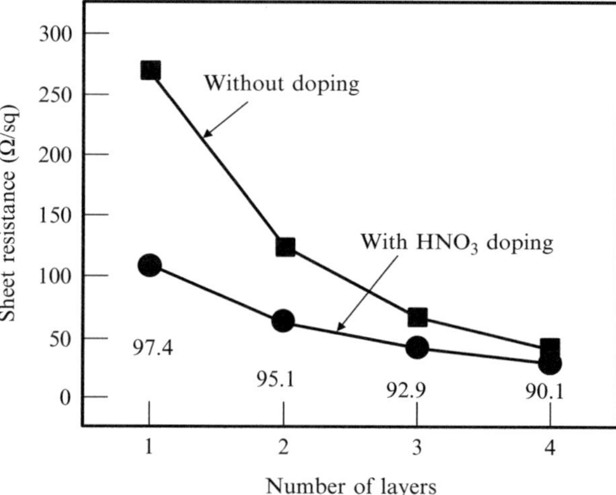

Fig. 3.23 Sheet resistance change as a function of layer number of CVD graphene sheets [38]. The *inset numbers* denote light transmittance at 550 nm

SWCNT bundle size on the sheet resistance–transparency properties [36]. These SWCNT networks can reach a sheet resistance of as low as 110 Ω/sq. at 90 % optical transmittance with suitable chemical doping.

Graphene, a two-dimensional allotrope of carbon, is another attractive carbon nanomaterial, though it is in the development stage. It is predicted that the sheet resistance of highly doped graphene will vary with the number of layers as $R_{sh} = 62.4/N$ (Ω/sq.), where N is the number of layers [37]. A single graphene is found to absorb a 2.3 % fraction of incident white light [38]. This means that the transmittance of graphene sheets will vary as $T = 100 - 2.3N$ (%), where N is again the number of layers. Figure 3.23 shows the sheet resistance change as a function of layer numbers of the chemical vapor deposition (CVD) fabricated graphene film on a substrate [39]. A doped four-layer graphene film exhibits a sheet resistance as low as around 30 Ω/sq. at 90 % optical transparency. Typical resistivity–transparency values are tabulated in Table 3.4 based on a recent review [40]. These results suggest that the CVD

Table 3.4 Resistivity and transparency for graphene at a wavelength of 550 nm [40]. "rGO" means reduced graphene oxide

Graphene material		Resistivity	Transparency (%)
Exfoliated graphite		5 KΩ/cm^2	90
		8 KΩ/cm^2	83
		1 K–1 MΩ/cm^2	30–90
Reduced graphene oxide		1.8 KΩ/cm^2	70
		1 KΩ/cm^2	80
		5 KΩ/cm^2	80
		70 KΩ/cm^2	65
		19 MΩ	95
		11 KΩ/cm^2	96
		1,425 S/cm	70
Graphene hybrid	rGO-Silica	0.45 S/cm	94
	rGO-CNT	240 Ω/cm^2	85
	rGO-CNT	151 KΩ/cm^2	93
CVD graphene		280 Ω/cm^2	80
		350 Ω/cm^2	90
		700 Ω/cm^2	80

graphene is as high quality as TCF and the graphene-CNT hybrid seems to be another very promising direction. In contrast, since CVD graphene is quite expensive, the reduced graphene hybrid with CNTs is expected to be a good choice as a cost-effective TCF.

In the last part of this section, metallic nanowires, i.e., Ag and Cu nanowires, are compared with other materials as TCF elements. The synthesis methods include a CVD method, electrochemical deposition, soft or hard template processes, and solution-chemical methods. As already mentioned at the beginning of this section, the performance of TCFs with metallic nanowires is superior to that of the others with PEDOT/PSS, oxides, SWCNTs, or graphene. The sheet resistivity can be held below 10 Ω cm while transparency exceeds 90 %. Figure 3.24 shows a typical network structures of Ag nanowires. However, as-printed Ag nanowire film only shows a very high resistivity of 10^4 Ω cm. After curing at 200 °C, the film exhibits low sheet resistance, 10 Ω cm. The high value of as-printed wires is caused by the presence of a thin layer on Ag nanowires. Due to the nanowire, synthesis requires PVP in a solution, and PVP remains on the surface of nanowires at a thickness of less than 10 nm. Even though an intensive washing treatment can lower the sintering temperature, complete removal of PVP is impossible. Achieving good electrical contact through a PVP layer requires high-temperature curing treatment.

Cu nanowires have also attracted the attention of researchers for many years. Of the Cu nanowires, the solution-chemical methods are likely to be the most suitable for mass production [41, 42]. Cu compounds such as $Cu(NO_3)_2$ or $CuCl_2$ were mixed with NaOH, EDA, and hydrazine and were reacted below 100 °C. When CuNWs follow a redox reaction:

$$2Cu^{2+} + N_2H_4 + 4OH \rightarrow 2Cu + N_2 + 4H_2O.$$

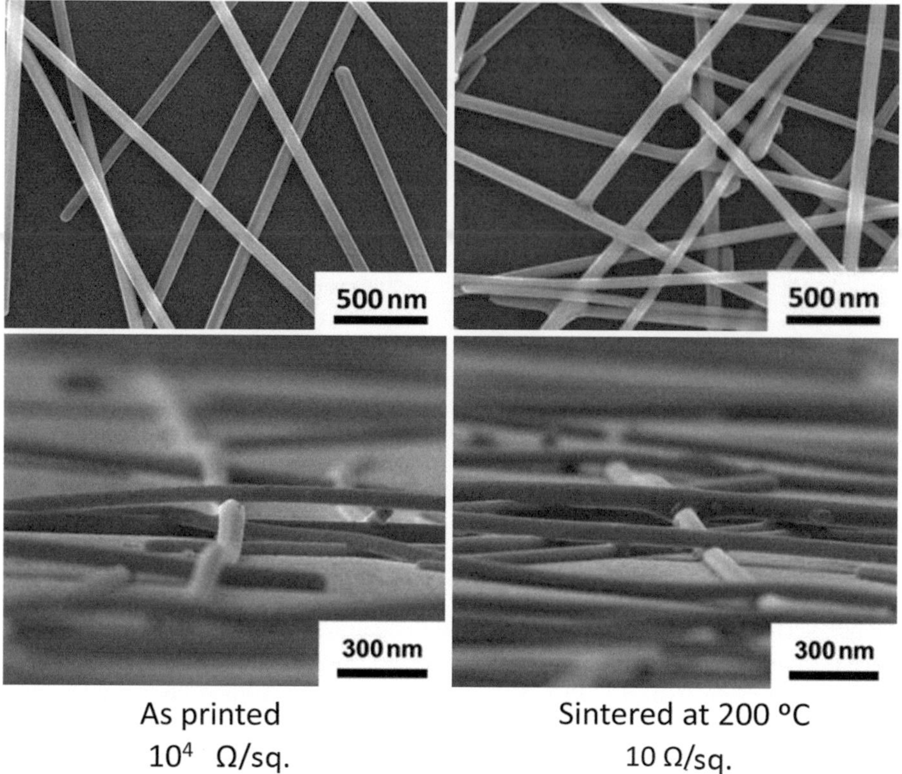

Fig. 3.24 Ag nanowire network formed on glass substrate [40]

As shown in Fig. 3.25, the obtained CuNWs are very long, up to 50 μm in length, and 60–160 nm thick [41]. The performance of the CuNW network on a flexible plastic film is expected to be similar to that with AgNWs since the bulk resistivity and morphology of both nanowires are similar. Since the oxidation susceptibility of CuNWs is much stronger than that of AgNWs, one needs to control the sintering atmosphere. To prevent oxidation effects, H_2 atm sintering was carried out. Figure 3.26 shows a typical transparency–sheet resistance curve for a CuNW network after sintering at 175 °C in H_2 atm [42]. An excellent TCF property is achieved for the CuNW network, which is equivalent to the AgNW network TCF.

3.7   Low Temperature Fabrication of Metal Nanowire TCF

To avoid the high-temperature treatment for TCF fabrication mentioned in the previous section, two possible methods have been proposed.

The first one involves the mechanical forming or pressing of nanowire films. Nanowires can plastically deform, and the deformation results in a complete

Fig. 3.25   (a) CuNWs in mother liquor and (b) SEM of CuNWs [40]. (Reprinted with permission from ref. [40]. Copyright (2013) American Chemical Society)

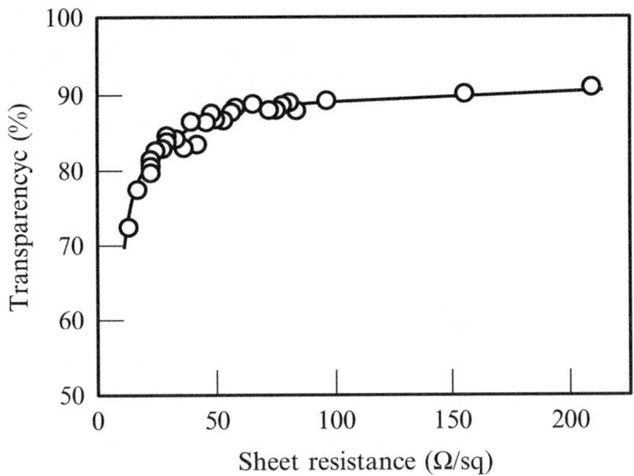

Fig. 3.26   Plot of transmittance (at a wavelength of 550 nm) vs. sheet resistance for films of CuNWs [41]

formation of metallic junctions among nanowires at room temperature. Figure 3.27 shows this effect for the formation of a AgNW network on a PET film [43]. Mechanical pressing provides two other features. First of all, it makes the network surface flat. In a solution printing process for nanowires, the roughness of the nanowire network tends to be greater due to the freedom of arrangement. As shown in Fig. 3.27, the pressing makes the flat surface of metallic nanowire network. For instance, the as-printed initial roughness of 560 nm becomes 120 nm after pressing,

Fig. 3.27 Cold-pressed Ag nanowire TCF on PET film [42]. The sheet resistance is 10 Ω/sq. with a light transmittance of more than 80 %

which is close to the nanowire thickness. Flatness is one of the essential requirements for organic devices, such as photovoltaic cells and OLED lighting, because a typical semiconductor layer thickness is less than 100 nm for those devices. The second benefit of metallic nanowires is that they form a perfect bond on a plastic substrate. Figure 3.25 shows the intimate contact between the AgNWs and the PET substrate. A bending fatigue test revealed no degradation for this AgNW network up to 1,000 bending cycles, while that formed by heating degrades by bending easily, as shown in Fig. 3.28.

Transparent paper with nanocellulose fibers is one of the most interesting substrates for PE technology [44] since it possesses excellent transparency like that of a PET film, low thermal expansion equivalent to that of glass, greater strength than steel, and an elastic modulus close to that of steel. The thickness can be controlled from a few microns to several hundred microns using a conventional papermaking process. No polymer binder is required. Above all, it is abundant and environmentally friendly because it is biodegradable and disposable. It was reported that the compatibility between AgNWs and transparent paper is perfect [45]. Figure 3.29 shows a TCF with a transparent nanocellulose fiber paper as substrate that was

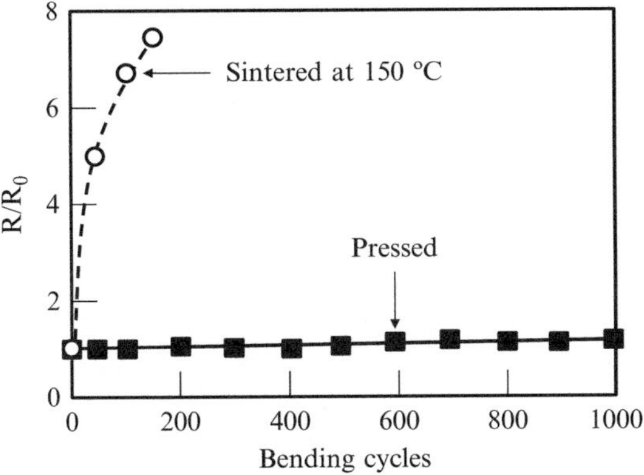

Fig. 3.28 Sheet resistance change as a function of bending cycles [42]. Bending radius was 5 mm. R/R_0 is ratio of a given sheet resistance to initial one

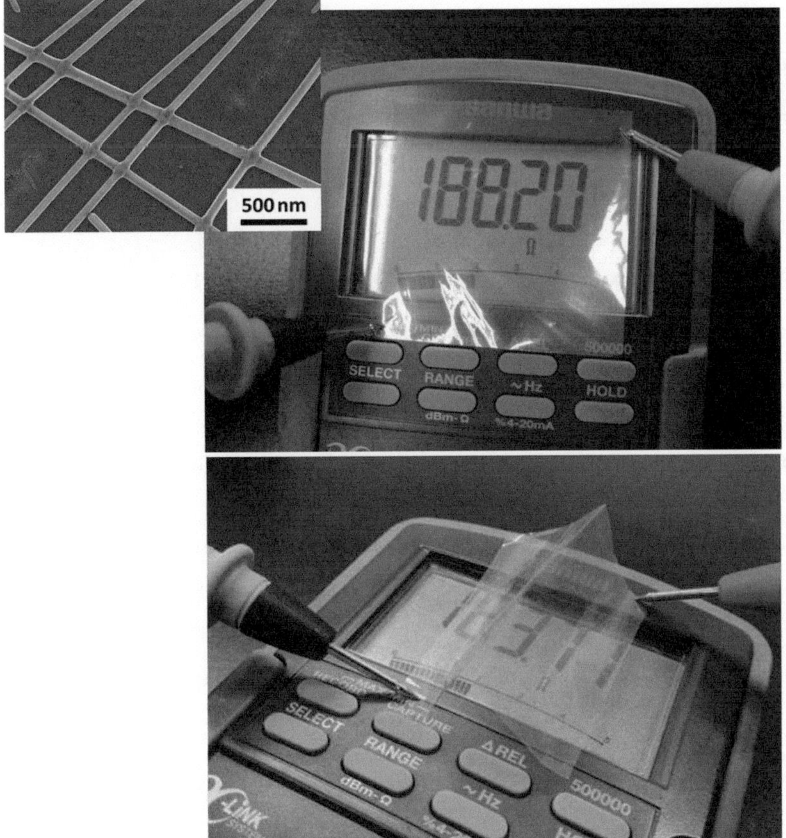

Fig. 3.29 AgNW TCF on nanocellulose fiber transparent paper formed by room-temperature pressing [43]. Complete holding does not influence the sheet resistance

press-formed at room temperature. The tester in the photograph is showing the sheet resistance before and after origami-like mountain fold the TCF. A plane TCF resistance, 188 Ω/sq., becomes 183 Ω/sq. after mountain holding. There is no difference within a margin of error, and mountain holding does not affect resistivity. This feature cannot be achieved for TCFs on a PET or any other flexible substrate. Thus, the affinity of AgNWs to nanocellulose transparent paper is excellent. A bending fatigue test of the AgNWs/nanocellulose paper TCF showed no effect of bending on its sheet resistance [45].

The second method is flash lamp sintering of nanowires on a plastic substrate. Flash lamp sintering was discussed in the previous chapter, and it is effective also for TCF fabrication.

References

1. Behabtu N et al (2013) Strong, light, multifunctional fibers of carbon nanotubes with ultrahigh conductivity. Science 339(6116):182–186
2. Buffat PH, Borel J-P (1976) Size effect on the melting temperature of gold particles. Phys Rev A 13(6):2287–2298
3. Brust M, Walker M, Bethell D, Schiffrin DJ, Whyman R (1994) Synthesis of thiol derivatized gold nanoparticles in a 2-phase liquid-liquid system. J Chem Soc Chem Commun 7:801–802
4. Daniel M-C, Astruc D (2004) Gold nanoparticles: Assembly, supramolecular chemistry, quantum-size-related properties, and applications toward biology, catalysis, and nanotechnology. Chem Rev 104:293–346
5. Yonezawa T, Yasui K, Kimizuka N (2001) Controlled formation of smaller gold nanoparticles by the use of four-chained disulfide stabilizer. Langmuir 17(2):271–273
6. Jadzinsky PD, Calero G, Ackerson CJ, Bushnell DA, Kornberg RD (2007) Structure of a thiol monolayer-protected gold nanoparticle at 1.1 Å resolution. Science 318(5849):430–433
7. Kanehara M, Takeya J, Uemura T, Murata H, Takimiya K, Sekine H, Teranishi T (2012) Electroconductive π-junction Au nanoparticles. Bull Chem Soc Jpn 85(9):957–961
8. Prabhu S, Poulose EK (2012) Silver nanoparticles: mechanism of antimicrobial action, synthesis, medical applications, and toxicity effects. Int Nano Lett 2:32
9. Albeniz AC, Barabera J, Espinet P, Lequerica MC, Levelut AM, Lopez-Marcos FJ, Serrano JL (2000) Ionic silver amino complexes displaying liquid crystalline behavior close to room temperature. Eur J Inorg Chem 2000(1):133–138
10. Wakuda D, Hatamura M, Suganuma K (2007) Novel method for room temperature sintering of Ag nanoparticle paste in air. Chem Phys Lett 441:305–308
11. Kawasaki H, Kosaka Y, Myoujin Y, Narushima T, Yonezawa T, Arakawa R (2011) Microwave-assisted polyol synthesis of copper nanocrystals without using additional protective agents. Chem Commun 47:7740–7742
12. Park BK, Kim DJ (2007) S.H Jeong, J.H. Moon, J.S. Kim, Direct writing of copper conductive patterns by ink-jet printing. Thin Solid Films 515:7706–7711
13. Grouchko M, Kamyshny A, Magdassi S (2009) Formation of air-stable copper–silver core–shell nanoparticles for inkjet printing. J Mater Chem 19(19):3057–3062
14. Teng KF, Vest RW (1987) Liquid ink jet printing with MOD inks for hybrid microcircuits. IEEE Trans Comp Hybrids Manu Technol 10(4):545–549
15. Cuk T, Troian SM, Hong CM, Wagner S (2000) Using convective flow splitting for the direct printing of fine copper lines. Appl Phys Lett 77:2063–2065
16. Dearden AL, Smith PJ, Shin D-Y, Reis N, Derby B, O'Brien P (2005) A low curing temperature silver ink for use in ink-jet printing and subsequent production of conductive tacks. Macromol Rapid Commun 26(4):315–318

17. Kawazome M, Suganuma K, Hatamura M, Kim K-S, Horie S, Hirasawa A, Tanaami H (2012) Low temperature printing wiring with Ag salt pastes. In: Proc. 39th International Symposium on Microelectronics (IMAPS2006), San Diego, October 8–12, pp.1050–1055
18. Hirose K, Kawazome M, Sekiguchi T, Hatamura M, Suganuma K (2012) Low temperature wiring technology with silver β-ketocarboxylate. IEICE Trans Electron J95-C(11):394–399
19. Araki T, Sugahara T, Jiu J, Nagao S, Nogi M, Koga H, Uchida H, Shinozaki K, Suganuma K (2013) Cu salt ink formulation for printed electronics using photonic sintering. Langmuir 29(35):11192–11197
20. Korte KE, Skrabalak SE, Xia Y (2008) Rapid synthesis of silver nanowires through a CuCl⁻ or CuCl₂-mediated polyol process. J Mater Chem 18:437–441
21. Jiu J, Murai K, Kim D, Kim K-S, Suganuma K (2009) Preparation of Ag nanorods with high yield by polyol process. Mater Chem Phys 114:333–338
22. Jiu J, Tokuno T, Nogi M, Suganuma K (2012) Synthesis and application of Ag nanowires via a trace salt assisted hydrothermal process. J Nanopart Res 14:975 (11pp)
23. Rathmell AR, Wiley BJ (2011) The synthesis and coating of long, thin copper nanowires to make flexible, transparent conducting films on plastic substrates. Adv Mater 23:4798–4803
24. Mohl M, Pusztai P, Kukovecz A, Konya Z (2010) Low-temperature large-scale synthesis and electrical testing of ultralong copper nanowires. Langmuir 26(21):16496–16502
25. Lofton C, Sigmund W (2005) Mechanism controlling crystal habits of gold and silver colloids. Adv Funct Mater 15:1197–1208
26. Sellmyer DJ, Zheng M, Skomski R (2001) Magnetism of Fe, Co and Ni nanowires in self-assembled arrays. J Phys Condens Matter 13:R433–R460
27. Shirakawa H, Louis EJ, MacDiarmid AG, Chiang CK, Heeger AJ (1977) Synthesis of electrically conducting organic polymers: Halogen derivatives of polyacetylene, (CH) x. J Chem Soc Chem Commun 16:578 580
28. Groenendaal LB, Jonas F, Freitag D, Pielartzik H, Reynolds JR (2000) Poly(3, 4-ethylene-dioxythiophene) and its derivatives: past, present, and future. Adv Mater 12(7):481–494
29. Tomonaga H, Morimoto T (2001) Indium-tin oxide coatings via chemical solution deposition. Thin Solid Films 392:243–248
30. Lin Y-H, Faber H, Zhao K, Wang Q, Amassian A, McLachlan M, Anthopoulos TD (2013) High- performance ZnO transistors processed via an aqueous carbon-free metal oxide precursor route at temperatures between 80–180 °C. Adv Mater 25:4340–4346
31. Kim Y-H, Heo J-S, Kim T-H, Park S, Yoon M-H, Kim J, Oh MS, Yi G-R, Noh Y-Y, Park SK (2012) Flexible metal-oxide devices made by room temperature photochemical activation of sol-gel films. Nature 489:128–132
32. Hosono H (2006) Ionic amorphous oxide semiconductors: material design, carrier transport, and device application. J Non-Cryst Solids 352:851–858
33. Nakamoto M (2009) Development of ITO nanoparticle ink for the transparent conductive film formation and the properties of the ITO film. J Surf Finish Soc Jpn 60(10):631–635
34. Osawa M, Yubashi S, Hayashi S, Oda M (2008) Development of transparent conductive film ink and its feature. ULVAC Tech J 68:24
35. Hu L, Wu H, Cui Y (2011) Metal nanogrids, nanowires, and nanofibers for transparent electrodes. MRS Bull 36:760–765
36. Wu JB, Agrawal M, Becerril HA, Bao ZN, Liu ZF, Chen YS, Peumans P (2010) Organic light-emitting diodes on solution-processed graphene transparent electrodes. ACS Nano 4:43–48
37. Nair RR, Blake P, Grigorenko AN, Novoselov KS, Booth TJ, Stauber T, Peres NMR, Geim AK (2008) Fine structure constant defines visual transparency of graphene. Science 320:1308
38. Bae S, Kim H, Lee Y, Xu X, Park J-S, Zheng Y, Balakrishnan J, Lei T, Kim HR, Song YI, Kim Y-J, Kim KS, Özyilmaz B, Ahn J-H, Hong BH, Iijima S (2010) Roll-to-roll production of 30-inch graphene films for transparent electrodes. Nat Nanotechnol 5:574–578
39. Wassei JK, Kaner RB (2010) Graphene, a promising transparent conductor. Mater Today 13(3):52–59
40. Chang Y, Lye ML, Zeng HC (2005) Large-scale synthesis of high-quality ultralong copper nanowires. Langmuir 21:3746–3748

41. Rathmell AR, Bergin SM, Hua Y-L, Li Z-Y, Wiley BJ (2010) The growth mechanism of copper nanowires and their properties in flexible, transparent conducting films. Adv Mater 22:3558–3563
42. Tokuno T, Nogi M, Karakawa M, Jiu J, Aso Y, Suganuma K (2011) Fabrication of silver nanowire transparent electrodes at room temperature. Nano Res 4:1215–1222
43. Nogi M, Iwamoto S, Nakagaito AN, Yano H (2009) Optically transparent nanofiber paper. Adv Mater 21(16):1595–1598
44. Nogi M, Komoda N, Otsuka K, Suganuma K (2013) Foldable nanopaper antennas for origami electronics. Nanoscale 5:4395–4399
45. Jiu J, Nogi M, Sugahara T, Tokuno T, Araki T, Komoda N, Suganuma K, Uchida H, Shinozaki K (2012) Strongly adhesive and flexible transparent silver nanowire conductive film fabricated with high-intensity pulsed light technique. J Mater Chem 22:23561–23567

Chapter 4
Semiconductor Materials

4.1 Material Category and Some History

Si, having a diamond crystalline structure, is an indispensable semiconductor and has been a major player as a solid-state semiconductor for electronics since 1950s. For PE technology, however, Si semiconductors should be replaced by new materials that can be formulated into inks or modified into Si inks with a certain low-temperature manufacturing process.

As a transistor material for PE technology, there are three key transistor metrics, i.e., charge (electron and hole) mobility, on/off ratio, and threshold voltage. Of these, charge mobility is the essential parameter for electronic devices. Table 4.1 compares a typical charge mobility of various PE materials with types of Si. The electron mobility of Si reaches 1,500 cm^2/V s depending on the doping, while hole mobility is less than 450 cm^2/V s. However, as can be seen in Table 4.1, amorphous Si has a low electron mobility of less than 1 cm^2/V s. Amorphous Si, which is processed by PVD or CVD methods, has been applied widely for many years for thin-film transistors (TFTs), photovoltaic cells, and many other devices. Despite its low electron mobility, active matrix TFTs of amorphous Si are the major back-plane driver of e-books. A liquid crystal display (LCD) requires greater mobility, over 10 cm^2/V s beyond that of amorphous Si. Currently, the fabrication of LCD back planes involves the low-temperature crystallization of amorphous Si. Since RIFD technology requires megahertz to gigahertz responses for devices, a higher mobility, greater than 50 cm^2/V s, will be the target. Computing devices, such as CPUs and GPUs, require much higher levels, and so currently it is not realistic to think that Si devices will be replaced by PE devices.

K. Suganuma, *Introduction to Printed Electronics*, SpringerBriefs in Electrical and Computer Engineering 74, DOI 10.1007/978-1-4614-9625-0_4,
© Springer Science+Business Media New York 2014

Table 4.1 Comparison of charge mobility for PE technology

Materials		Mobility (cm^2/V s)	Process temperature (°C)	Notes
Organic		1–40	60–200	p-type, transparent
		0.1–5		n-type, transparent
Oxides		1–100	180–500	Transparent
Si	Nanoparticle	Approx. 100	~400	Too high temperature
	Polysilane	100–400	~400	Too high temperature
Si		1,000–1,500	–	Single crystal
		10–500	–	Polycrystal (crystallized at low/ high temperatures), transparent
		Approx. 1	–	Amorphous, transparent

4.2   Organic Semiconductors

In the mid-1980s, the initial electron mobility and on/off ratio of organic transistors were on the order of 10^{-5} cm^2/V s and 10^2, respectively. These values now exceed 10 cm^2/V s and 10^6, respectively. Figure 4.1 shows the historic improvements in the organic transistor properties from the literature. The basic molecular structures consist of thiophene and benzene, which have conjugated π-orbitals. Some of the organic TFT materials that are currently being used are shown in Fig. 4.2. It is interesting to note that the electron mobility of both thiophene and pentacene tended to saturate to a few square centimeters per/Volt-second. Those TFTs were mainly fabricated by vapor processes. Suddenly the electron mobility started to increase after the finding that a single crystal rubrene exhibited high mobility above 5 cm^2/V s [1]. At the time of this breakthrough, organic electronics was naturally merging into printing technology. Since printing technology requires stable ink material in ambient atmosphere, it was found that unstable chemicals in air might have some difficulty in industrial-scale applications. In this respect, pentacene has an instability problem, even though it has been recognized as one of the standard organic electronic materials. Instead of pentacene, organic materials compatible with printing technology have been intensively sought out. Takimiya proposed a new compound, [1]benzothieno[3,2-b]benzothiophene (C_n-BTBT), that had excellent stability in air and solubility [2]. This result indicates that small molecules possessing an extended aromatic core with solubilizing long aliphatic chains are promising candidates for printing technology.

The intrinsic carrier mobility of electrons (or holes) inside a single conjugated molecule is believed to be quite large. The maximum transistor mobility inside a single rubrene molecule is very high. In fact, that of measured in-crystal carriers reaches 18 cm^2/V s, which is usually influenced by several factors, such as purity/ crystalline quality, contact resistance, transistor configuration, substrate materials, and others. The contact-free intrinsic value is estimated to be 40 cm^2/V s, and thus the true mobility of a single rubrene molecule is equivalent to or greater than 40 cm^2/V s [3].

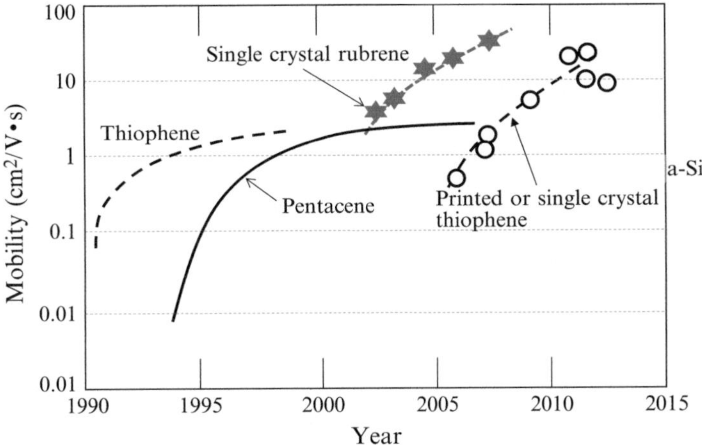

Fig. 4.1 Improvement of charge mobility of p-channel molecules over 2 decades

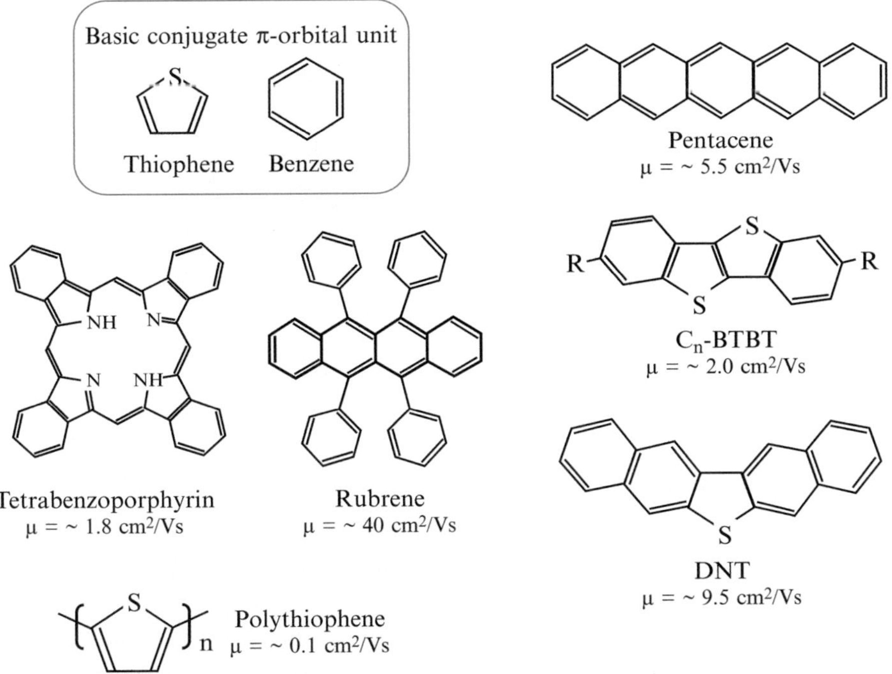

Fig. 4.2 Typical organic semiconductor and charge mobility

Organic field effect transistors (OFETs) are strongly determined by the morphology of the thin films and by the packing and orientation between individual molecules. Specifically, molecular packing strongly impacts the electronic coupling among molecules and the resulting charge carrier mobility. Figure 4.3 shows a

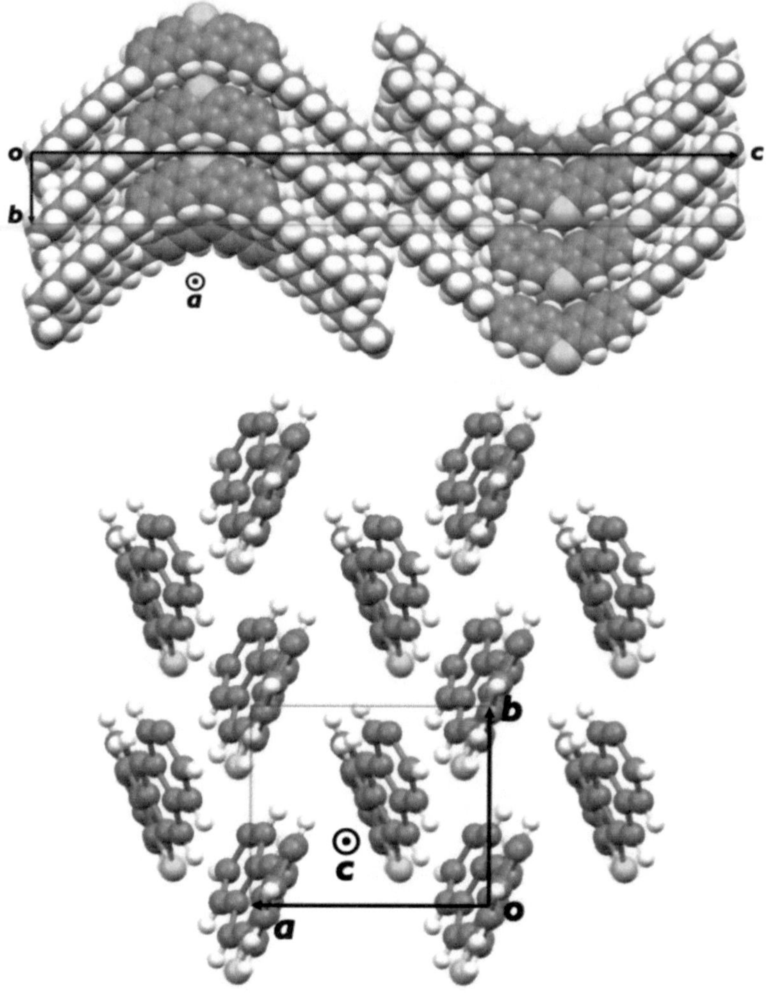

Fig. 4.3 Packing structures of C10–DNT–VV (alkyl chain omitted) [4]

sample packing of a molecule of C_{10}-DNT, which has a high mobility, 9.5 cm^2/V s, and was recently developed [4]. Unlike with Si or other inorganic semiconductors whose charge transfer, both of electrons and holes, is governed by a band transport mechanism, the charge transfer of OFETs is primarily governed by a bandlike transfer localized inside a single molecule and a hopping among neighboring molecules. The intermolecular distance in molecular packing plays an important role in hopping transfer. One must understand the difference between atoms in inorganic semiconductors, in which atoms are strictly bound inside a lattice, and organic molecules, which can vibrate or rotate rather freely. Free molecular motion results in a distortion of the hopping charge transfer. In other words, if the packing of molecules

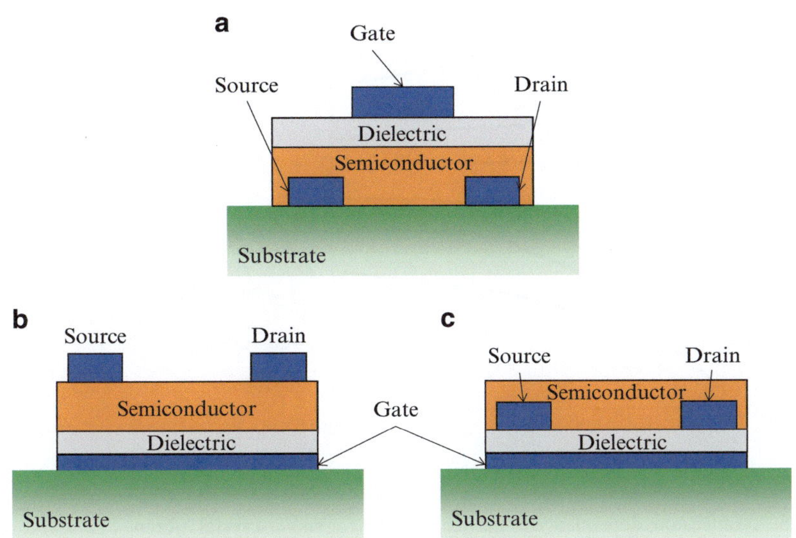

Fig. 4.4 Typical OFET structure: (**a**) top-gate type, (**b**) top-contact type, and (**c**) bottom-contact type

becomes tight in a single crystal, the charge transfer can be governed by a band mechanism. In fact, a bandlike transport mechanism was confirmed by a Hall effect measurement [5].

To date, many types of OFET structures have been developed and proposed to improve semiconductor performance. Figure 4.4 shows three representative OFET structures. Atmospheric temperature may have some influence since the thermal vibration or rotation of molecules retards charge hopping. Figure 4.5 shows the influence of temperature on the charge mobility of OFETs [6]. It is obvious that the temperature dependence of charge mobility is significant and apparently changes from negative to positive when changing substrates. Charges move near the surface of an organic semiconductor at the interface with a dielectric layer, and this nearby dielectric layer also has a strong influence whose mechanism is not fully understood. Figure 4.5 also shows a strong influence of dielectric materials on mobility. Mobility tends to decrease with an increasing gate insulator dielectric constant, except for Si_3N_4, which is thought to be a Si–O–N compound. Such a gate material influence can be attributed to the coupling effect between the carriers in an organic semiconductor and the opposite carriers in the nearby gate dielectric layer, as schematically shown in the figure. For sufficiently large coupling, which is the case with a larger dielectric constant, the hole and induced charge can move together to significantly affect device performance, which is quite different from that of inorganic semiconductors.

To understand the true electronic behavior of OFETs, one needs to use robust measurement methods to eliminate all artifacts. For instance, conductivity measurements using a two-probe OFET configuration comprises two contacts and a gate electrode. One must ensure that the apparent mobility determined in a two-probe

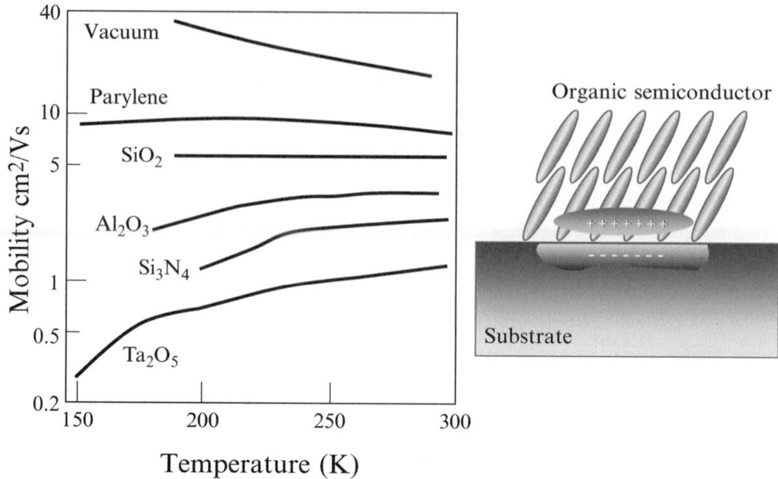

Fig. 4.5 Temperature dependence of carrier mobility for single-crystal rubrene FETs with six different gate dielectrics [4]. Dielectric constant: vacuum ($\varepsilon = 1$), parylene($\varepsilon = 2.9$), SiO$_2$ ($\varepsilon = 3.9$), Si$_3$N$_4$ ($\varepsilon = 7.5$), Al$_2$O$_3$ ($\varepsilon = 9.4$), and Ta$_2$O$_5$ ($\varepsilon = 25$)

measurement is not significantly affected by the parasitic contact resistance. But if the contact resistance is unknown, it is usually not negligible, and a four-probe measurement configuration must be used to separately measure the channel and the contact resistances.

For many years, it has been believed that OFETs have a low carrier mobility, except for single crystals. Naturally, the performance of a solution-processed OFET array was believed not to achieve a high mobility. But, fortunately, this turned out not to be true. Many patterning methods have been surveyed to improve OFET performance, and several reports showed the potential of specific patterning methods [7, 8]. Figure 4.6 shows a schematic of the edge cast method for matrix patterning [7]. The stamp has special inclined edges. Along the edges, the evaporation of solvent moving from the wider sides to the narrower sides accompanies the unidirectional crystallization of an organic material, resulting in the formation of a single crystal array. The average carrier mobility of C10-DNTT is approximately 15 cm^2/V s. Figure 4.6 shows the world's first LCD panel driven by an organic thin film transistor (OTFT). Similarly, an inkjet printing with two droplets, where one droplet serves as substrate base and the other serves as surface liquid solidified unidirectionally on the first droplet whose highest carrier mobility exceeds 30 cm^2/V s [8]. Thus, a directional solidification is one of the promising printing processes to achieve high-performance OFETs.

There are two main technological deficiencies in the fabrication of OTFTs. One is homogeneity and the other is the development of n-type transistors. Currently, printed single crystalline OTFT arrays have relatively large scatter when making printed patters. Purification and their single crystallinity significantly influence scatter. Compared with high-performance p-type OTFTs, the performance of n-type organic

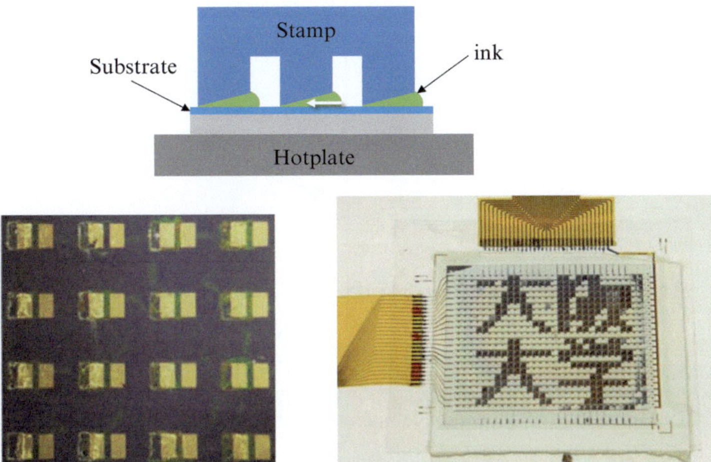

Fig. 4.6 Edge casting method for OTFT array with C10-DNTT and its LCD application (Courtesy of Prof. Takeya, University of Tokyo) [7]. The *white arrow* indicates the crystal growth direction

transistors is relatively low. For example, the charge mobility of an n-type semiconductor is one order lower than that of a p-type semiconductor. Since CMOS devices require the combination of p- and n-type transistors, it is challenging to realize high-performance OTFT with n-type transistors. There are, however, reports on the development of high-performance n-type transistors. Recently, electron mobility as high as 1–2 cm^2/V s and on/off current ratios greater than 10^6 were demonstrated [9, 10].

4.3 Oxide Semiconductors

Among ceramic materials, many oxides can transport carriers by the introduction of oxygen defects in lattices; this practice dates back to the 1960s. SnO_2 was proposed as a transparent oxide TFT comprised of an evaporated SnO_2 semiconductor on a glass substrate, and possessed a high mobility that reached 70 cm^2/V s in 1964 [11]. Due to their excellent performance (Table 4.1), oxide ceramics possess much higher carrier mobility than OTFTs and are basically transparent, as are OTFTs. Because of these attractive features, oxide transparent thin-film transistors have garnered much attention during the past decade. In particular, amorphous oxides represent promising TFT materials and have already shown impressive progress, particularly in display applications, in a relatively short period. Usually, for PE technology, oxide ceramics require a high process temperature for sintering and annealing, above 400 or 500 °C. The recent drop in process temperatures, however, reflected amazing progress for n-type oxide semiconductors.

Today, ZnO-based transparent TFTs have attracted major interest as transparent devices. Although the early reports on ZnO semiconductors showed poor

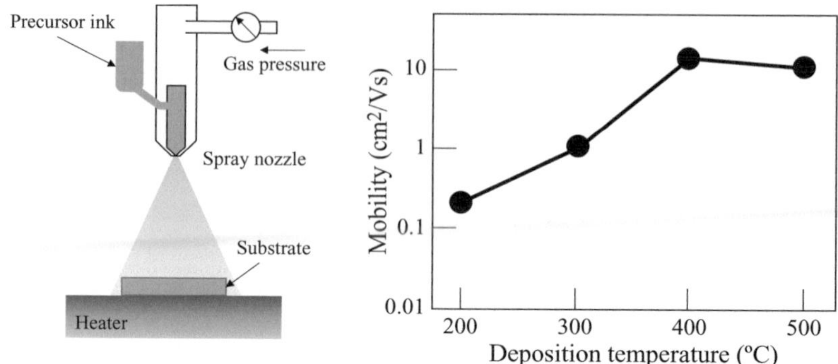

Fig. 4.7 Schematic representation of spray deposition apparatus and electron mobility of bottom-gate ZnO TFT as a function of deposition temperature [15]

performance, more advanced performance of ZnO TFTs were reported in the early 2000s [12, 13]. Hoffman reported that ZnO could achieve 75 % transparency in the visible spectrum with an on/off ratio of 10^7, and the mobility of ZnO-based devices ranges from 0.3 to 2.5 cm^2/V s [12]. An In–Ga–Zn–O system (amorphous-IGZO) was reported by Hosono's group in 2004 [13]. They showed that the a-IGZO deposited on a PET film at room temperature exhibited mobility over 10 cm^2/V s, which is much larger than a-Si, while a single-crystal IGZO exhibited a much higher mobility of 80 cm^2/V s [14]. Following the publication of these works, the great potential of oxide semiconductors has attracted the attention of many researchers.

As mentioned in the previous chapter, oxide films can be formed on a substrate by two typical ink forms, a sol-gel ink and a nanoparticle ink. The initial sol-gel work, in which $Zn(OCOCH_3)_2 \cdot 2H_2O$ was used as the starting precursor, has already demonstrated the excellent performance of ZnO after heat treatment at 600 or 900 °C [15]. Anthopoulos et al. demonstrated the use of a simple deposition technique for the fabrication of high-performance ZnO TFTs [16]. A spray deposition method was used, as shown in Fig. 4.7. Using this method, a high mobility of 15 cm^2/V s was achieved in a bottom-gate-type ZnO TFT fabricated at 400 °C. This process temperature is, however, still too high for most PE products.

Recent works have successfully employed lower-temperature processes. The Kanatzidis and Marks group demonstrated a new process, in which they obtained TFTs of In_2O_3, a-Zn–Sn–O, and a-In–Zn–O, as well as ITO, at low process temperatures by a self-combustion reaction during annealing [17] (Fig. 4.8). In particular, the In_2O_3 TFT exhibited excellent mobility when it was sintered at temperatures as low as 200 °C. Yang et al. fabricated a bottom-gate IGZO TFT by spin coating nanoparticles, followed by postbaking at a low temperature, 95 °C. The TFT device showed a charge mobility of 2.3 cm^2/V s and an on/off current ratio greater than 10^6 [18]. To obtain a high-performance TFT, a dense TFT structure formed at low temperature is essential. Deep-ultraviolet (deep-UV) irradiation also works well to produce efficient condensation and densification of oxide semiconductors by

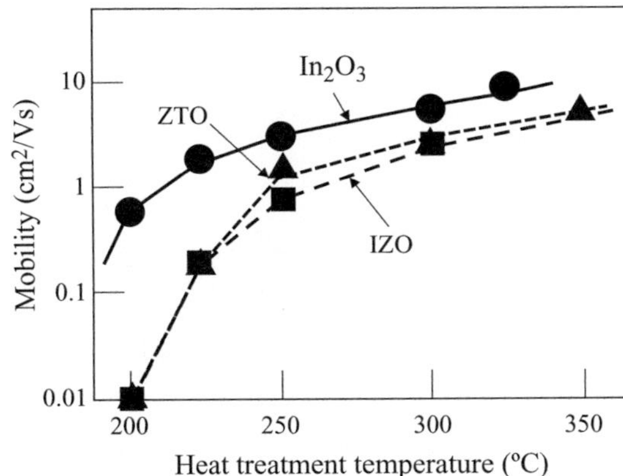

Fig. 4.8 Mobility change of TFTs inkjet printed on plastic substrate as a function of heat-treatment temperature [17]

photochemical activation [19]. The charge mobility of IGZO processed by deep-UV irradiation without additional heating was 3.77 cm^2/V s with a narrow scatter.

In contrast to the great advancements in n-type oxide TFTs, the reported performance of p-type oxide TFTs remains low; in particular, there is still no p-type oxide ink for use with PE technology. In other words, the combination of a p-type organic semiconductor and an n-type oxide one may be the best approach to assembling printed transparent TFT devices.

4.4   Other Semiconductors

Other potential materials for PE ink technology include Si nanoparticles and Si precursor inks. Even though the printing of such Si nanomaterials is a straightforward method of fabricating semiconductors, developments are still in the early stages. The two main challenges in this area are controlling the purity of nanoparticles or precursors and lowering the process temperature.

Figure 4.9 shows a bottom-gate TFT fabricated by screen printing without any postprocessing steps, such as sintering or calendaring [20]. The achieved on/off ratio, threshold voltage, and charge mobility were 600, −1.8 V, and 0.7 cm^2/V s, respectively. Such a performance is comparable to that of a-Si.

It is well known that polysilane, i.e., straight-chain (Si_nH_{2n+2}) or cyclic (Si_nH_{2n}), can be a precursor of Si by a decomposition reaction. Shimoda et al. demonstrated the potential of polycrystalline Si fabrication with a silane-based liquid precursor [21]. They utilized spin coating and inkjet printing to create a silane pattern. Sintering at 400 °C for 30 min was first carried out, and the temperature was subsequently increased to 540 °C for 2 h to form a 50-nm-thick a-Si film. The a-Si film was irradiated by an excimer laser to obtain polycrystalline Si TFTs, whose highest mobility was 108 cm^2/V s (Fig. 4.10).

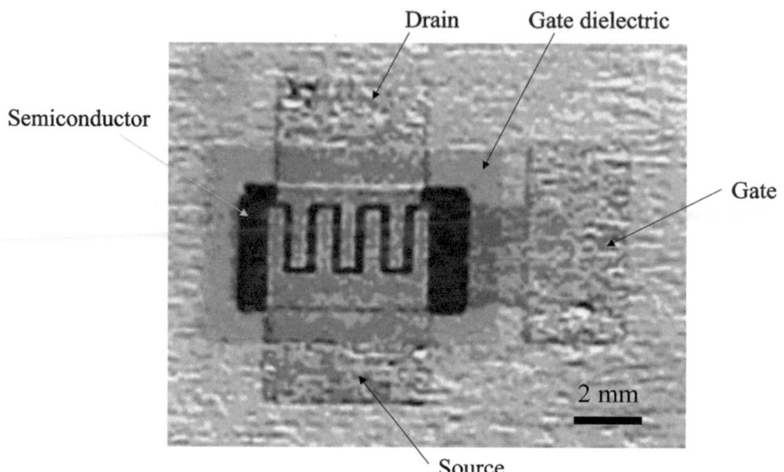

Fig. 4.9 Field effect transistor printed with Si nanoparticles on paper [20]. The gate, source, and drain are patterned with Ag nanoparticle ink and the gate length and width are 200 μm and 14.4 μm, respectively

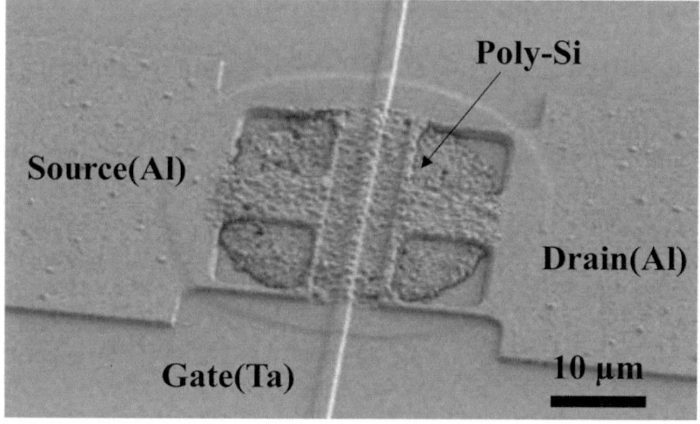

Fig. 4.10 SEM image of TFT made from inkjetted Si film [21]

References

1. Podzorov V, Sysoev SE, Loginova E, Pudalov VM, Gershenson ME (2003) Appl Phys Lett 83:3504
2. Ebata H, Izawa T, Miyazaki E, Takimiya K, Ikeda M, Kuwabara H, Yui T (2007) Highly soluble [1]benzothieno[3,2-b]benzothiophene (BTBT) derivatives for high-performance, solution-processed organic field-effect transistors. J Am Chem Soc 129:15732–15733
3. Takeya J, Yamagishi M, Tominari Y, Hirahara R, Nakazawa Y, Nishikawa T, Kawase T, Shimoda T, Ogawa S (2007) Very high-mobility organic single-crystal transistors with in-crystal conduction channels. Appl Phys Lett 90:102120

4. Okamoto T, Mitsui C, Yamagishi M, Nakahara K, Soeda J, Hirose Y, Miwa K, Sato H, Yamano A, Matsushita T, Uemura T, Takeya J (2013) V-shaped organic semiconductors with solution processability, high mobility, and high thermal durability. Adv Mater 25(44):6392–6397
5. Takeya J, Tsukagoshi K, Aoyagi Y, Takenobu T, Iwase Y (2005) Hall effect of quasi-hole gas in organic single-crystal transistors. Jpn J Appl Phys 44(46):L1393–L1396
6. Hulea IN, Fratini S, Xie H, Mulder CL, Iossad NN, Rastelli G, Ciuchi S, Morpurgo AF (2006) Tunable Frohlich polarons in organic single-crystal transistors. Nat Mater 5:982–986
7. Nakayama K, Hirose Y, Soeda J, Yoshizumi M, Uemura T, Uno M, Li W, Kang MJ, Yamagishi M, Okada Y, Miyazaki E, Nakazawa Y, Nakao A, Takimiya K, Takeya J (2011) Patternable solution-crystallized organic transistors with high charge carrier mobility. Adv Mater 23: 1626–1629
8. Minemawari H, Yamada T, Matsui H, Tsutsumi J, Haas S, Chiba R, Kumai R, Hasegawa T (2011) Inkjet printing of single-crystal films. Nature 475:364–367
9. Soeda J, Uemura T, Mizuno Y, Nakao A, Nakazawa Y, Facchetti A, Takeya J (2011) High electron mobility in air for N, N′-1H,1H-perfluorobutyldicyanoperylene carboxy-di-imide solution-crystallized thin-film transistors on hydrophobic surfaces. Adv Mater 23:3681–3685
10. Yun SW, Kim JH, Shin S, Yang H, An B-K, Yang L, Park SY (2012) High-performance n-type organic semiconductors: incorporating specific electron-withdrawing motifs to achieve tight molecular stacking and optimized energy levels. Adv Mater 24:911–915
11. Klasens HA, Koelmansa H (1964) A tin oxide field-effect transistor. Solid-State Electron 7:701–702
12. Hoffman RL, Norris BJ, Wager JF (2003) ZnO-based transparent thin-film transistors. Appl Phys Lett 82:733–735
13. Nomura K, Ohta H, Takagi A, Kamiya T, Hirano M, Hosono H (2004) Nature 432:488
14. Nomura K, Ohta H, Ueda K, Kamiya T, Hirano M, Hosono H (2003) Thin-film transistor fabricated in single-crystalline transparent oxide semiconductor. Science 300(5623):1269–1272
15. Ohya Y, Niwa T, Ban T, Takahashi Y (2001) Thin film transistor of ZnO fabricated by chemical solution deposition. Jpn J Appl Phys 40:297–298
16. Bashir A, Wobkenberg PH, Smith J, Ball JM, Adamopoulos G, Bradley DDC, Anthopoulos TD (2009) High-performance zinc oxide transistors and circuits fabricated by spray pyrolysis in ambient atmosphere. Adv Mater 21:2226–2231
17. Kim M-G, Kanatzidis MG, Facchetti A, Marks TJ (2011) Low-temperature fabrication of high-performance metal oxide thin-film electronics via combustion processing. Nat Mater 10:382–388
18. Yang Y-H, Yang SS, Kao C-Y, Chou K-S (2010) Chemical and electrical properties of low-temperature solution-processed In-Ga-Zn-O thin-film transistors. IEEE Eelectron Device Lett 31(4):329–331
19. Kim Y-H, Heo J-S, Kim T-H, Park S, Yoon M-H, Kim J, Oh MS, Yi G-R, Noh Y-Y, Park SK (2012) Flexible metal-oxide devices made by room temperature photochemical activation of sol–gel films. Nature 489:128–132
20. Härting M, Zhang J, Gamota DR, Britton DT (2009) Fully printed silicon field effect transistors. Appl Phys Lett 94:193509
21. Shimoda T, Matsuki Y, Furusawa M, Aoki T, Yudasaka I, Tanaka H, Iwasawa H, Wang D, Miyasaka M, Takeuchi Y (2006) Solution-processed silicon films and transistors. Nature 440:783–786

Chapter 5
Substrate and Barrier Film

5.1 Substrate

Printed electronics also requires advanced materials technology for substrates. The requirements include flexibility, excellent transparency in many cases, surface smoothness, thin and light weight, low thermal expansion, stiffness, heat resistance, low cost, and others. Several choices are available as substrates depending on the nature of the PE product. Of course, it is possible not to make any choice of substrate but direct printing on the cases of the equipment with a 3D printing technology. But using certain flexible substrates will be the majority for mass production and ultimately will be applied to a roll-to-roll process. Table 5.1 lists materials commonly used in PE products.

A glass substrate is an attractive transparent substrate and has been widely used for most optical purposes such as displays, photovoltaics, and lightings. Its transparency is greater than 90 %, with a Haze far below 1 %. Thin glass substrates are now commercially available, as shown in Fig. 5.1. The thickness of glass substrates has already reached 30 μm. In contrast, the weakness of glass substrates is their brittleness, heavy weight compared with plastics, and high cost. To provide flexibility and robustness to PE products, plastic, paper, or steel substrates are better.

PET (polyethylene terephthalate) film is the most popular and widely used plastic film. It has high optical transparency above 90 % and the great benefit of low cost compared with other substrates. One of the drawbacks of PET is its poor heat resistance. Due to its poor heat capability, entire printing process must be performed at low temperatures below 130 °C under low tension. The heat resistance gets better for polyethylene naphthalate (PEN) and much better for polyimide (PI) while transparency decreases and cost increases. Though polymeric films are the first choice for PE products, several things must be carefully controlled to minimize distortion, especially for roll-to-roll printing. These are the controls of atmosphere (temperature and humidity), tension, and drying, and distortion detection and high-resolution imaging in high-speed web handling.

K. Suganuma, *Introduction to Printed Electronics*, SpringerBriefs in Electrical and Computer Engineering 74, DOI 10.1007/978-1-4614-9625-0_5, © Springer Science+Business Media New York 2014

Table 5.1 Variety of flexible substrates and selected properties

	Thickness (µm)	Density (g/cm²)	Transparency (%)	Haze (%)	Tg (°C)	Process temperature limit (°C)	Notes
PET	16–100	1.4	90	Approx. 0.3	80	120	
PEN	12–250	1.4	87	Approx. 0.8	120	155	
PI	12–125	1.4	–	–	410	300	
Glass	50–700	2.5	90	0.1	500	400	
Paper	100	0.6–1.0	–	–	–	130	
Transparent paper	20–200	Approx. 1	90	1–2	200 (?)	150	Made of nanocellulose fibers [1]
Steel	200	7.9	–	–	–	600[a]	

[a]Needed to prevent oxidation

Fig. 5.1 A thin flexible glass substrate (Courtesy of Nippon Electric Glass, Osaka, Japan)

Paper substrates are also attractive for PE technology. The major advantage of paper substrates is their low cost and disposability due to their biodegradability. Disposable devices, such as RFIDs on a paper substrate without a Si chip, are expected to establish one of the biggest PE application markets. Paper without any polymer addition, which occurs nature, will be the main biodegradable substrate. In addition, transparent nanocellulose paper can provide display windows for paper devices [1, 2]. Nanocellulose paper possesses an excellent optical property (Fig. 5.2) [1]. The transparency of nanocellulose paper is almost equivalent to that of PET films and its heat resistance is similar to that of PEN, as shown in Figs. 5.3 and 5.4, respectively. The other attractive feature of transparent nanocellulose paper is its low thermal expansion and high Young's modulus. The thermal expansion coefficient is close to that of glass, $8 \times 10^{-6}/°C$, and the Young's modulus is

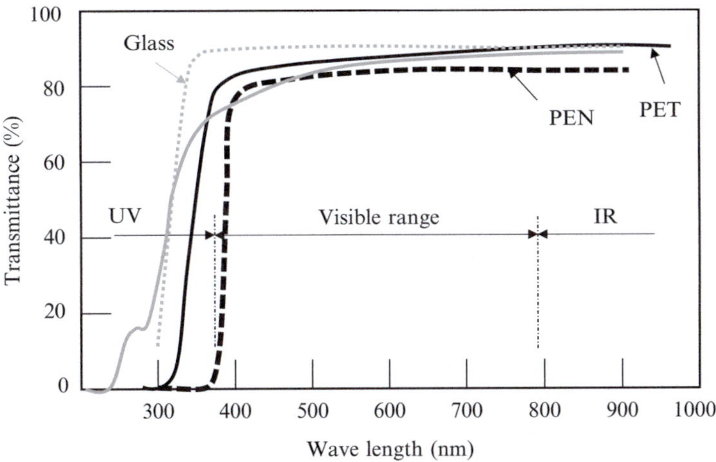

Fig. 5.2 Typical transmittance of transparent films as a function of wavelength. The thicknesses are 100 µm for PET and PEN and 750 µm for soda lime glass

Fig. 5.3 Transparent nanocellulose paper [1]

approximately140 GPa. These mechanical properties provide excellent dimensional stability.

Metallic substrates give PE applications heat resistant qualities. Steels, i.e., ferritic stainless and low-carbon steels, are typical substrates for photovoltaic cells printed with CIGS or with Si inks. Figure 5.5 shows a flexible stainless steel sheet

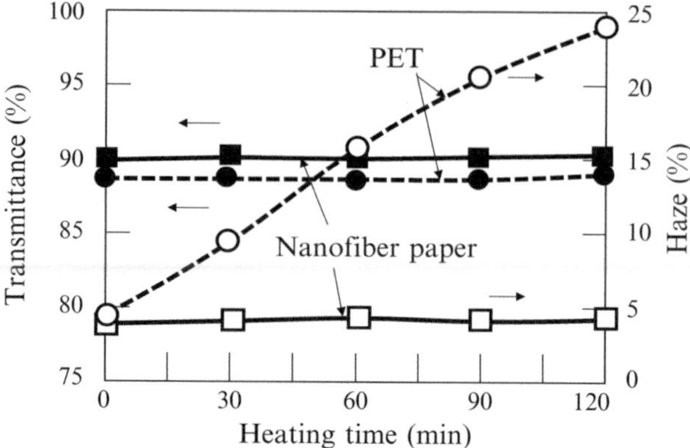

Fig. 5.4 Transparency of nanofiber paper and PET films [1]. Total light transmittance during heating at 150 °C for 120 min (*solid square*: nanofiber paper; *solid circles*: PET) and haze during heating at 150 °C for 120 min (*squares*: nanofiber paper; *circles*: PET)

Fig. 5.5 Stainless steel sheet for PE technology (Courtesy of Nippon Kinzoku, Tokyo, Japan)

roll. Its heat resistance is excellent up to 600 °C. It has a low thermal expansion of approximately 13×10^{-6}/°C and a Young's modulus as high as 200 GPa. These mechanical properties provide dimensional stability. Since metallic substrates are conductive, passivation treatment is required for PE substrates. The passivation layer should also have a diffusion barrier, especially for active elements such as Cr. On the other hand, the heavy weight of metallic substrates will limit their application.

Thus, since many options are available for substrates in PE applications, manufacturers must take these features into account to select the most suitable substrate for their own products.

5.2   Barrier Film Technology

Although there is no doubt that organic electronics are potential candidates for use in PE applications such as large-area displays, photovoltaics, lighting, or healthcare devices, the environmental stability and reliability of organic electronic devices are still major concerns. The degradation of organic devices can be attributed to various mechanisms, including crystallization of organic solids, electrochemical reactions at the heterointerfaces, and ion migration. It is well understood that organic electronics can be degraded in the presence of moisture. Not only organic devices, but metallic components such as Ag and Cu wiring can be severely degraded in moisture, as discussed in the following chapter.

Figure 5.6 shows a typical image of dark spots appearing on an OLED panel operated under moist conditions [3]. The panel has an Al layer as the backside cathode. The fact that these dark spots stop growing immediately when the driving current is shut down implies that the observed degradation is caused by an electrochemical process related to water. The electrochemical reduction of water, leading to the evolution of hydrogen gas, proceeds according to the following equation:

$$2H_2O + 2e^- \rightarrow H_2 + 2OH^-, \tag{5.1}$$

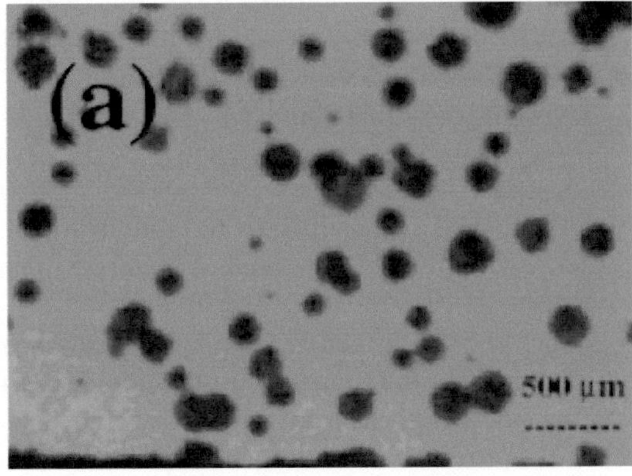

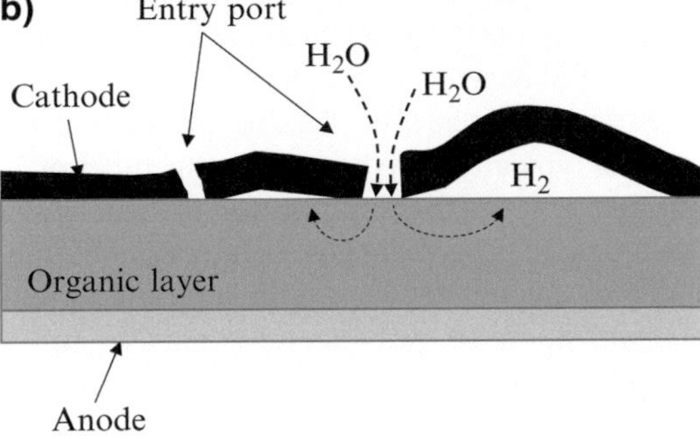

Fig. 5.6 (a) Optical microscope Organic layer image of dark spot on an OLED working at a luminance of 100 cd/m² in water vapor atmosphere and (b) schematic illustration of the degradation [3]

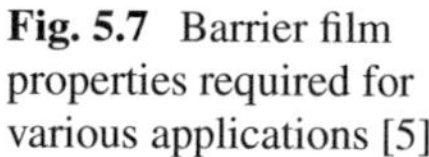

Fig. 5.7 Barrier film properties required for various applications [5]

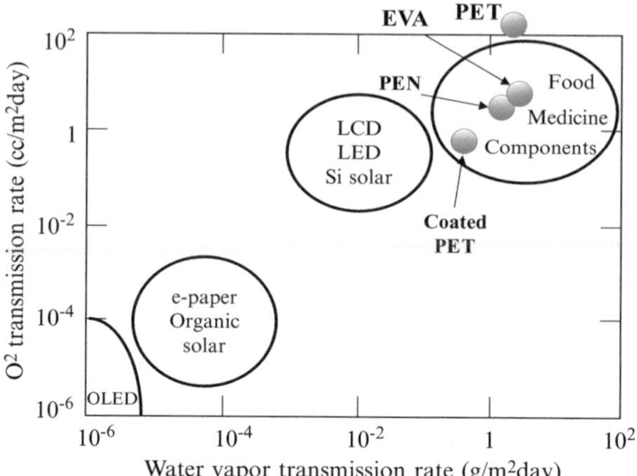

where the reduction of water occurs at a redox potential of −0.82 V. The degradation mechanism is schematically illustrated in Fig. 5.6b. There is another mechanism of dark spot formation related to OLEDs: the crystallization of OLEDs resulting in the formation of small protrusions under a local high electric field [4]. This causes delamination between the electrode and an OLED layer, creating nonemissive dark spots. The requirements for OLED displays are particularly high. The development of high barrier technology led to the production of many devices applicable to a wide spread area mentioned in Chapter 1.

Thus, barrier film technology is indispensable for protecting organic devices and metallic wiring for mass PE production. As a barrier layer, a glass substrate is perfect because it can prevent water vapor and oxygen transmission into devices. Nevertheless, glass has its own limitations to widespread application, as mentioned earlier. One of the key properties a substrate that might replace glass should possess would be a glasslike barrier property.

Generally, the attractive feature of plastic films as substrates and overlayers can be increased by low outgassing, low attack nature to devices/metallic wiring, low voiding and adhesion ability in addition to a high barrier property. Figure 5.7 depicts typical requirements for oxygen transmission and water vapor transmission for various organic devices with plastic barrier film properties [5]. As seen in the figure, high barrier performance, defined as a lower transmission of water vapor, less than 10^{-4} g/m²/day, and oxygen, less than 10^{-4} cc/m²/day, are required for many organic electronic devices. The flexible plastic substrates typically have barriers on the order of typically have barriers on the order of 1–10 g/m²/day for water vapor transmission and 1–10 cc/m²/day for oxygen transmission. No plastic satisfies these requirements. A typical barrier structure is shown schematically in Fig. 5.8.

There are several ways to give plastic films a high barrier performance, such as crystallization of polymers by tension forming, organic/microflake inorganic hybridization, and organic–inorganic multilayer coating. Of these, the incorporation of inorganic coating layers is the most promising. Of course, if an ITO film is

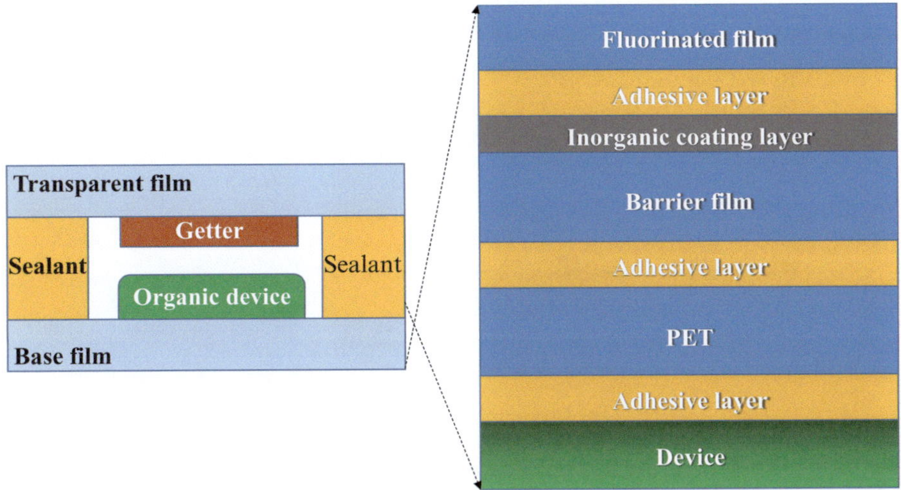

Fig. 5.8 Typical barrier layer structure for organic devices

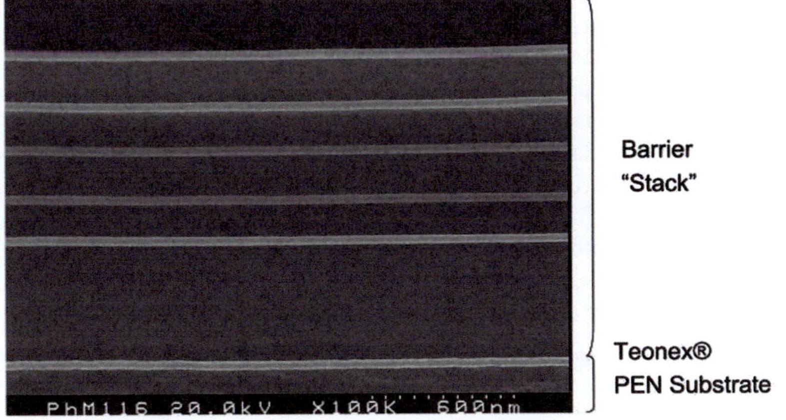

Fig. 5.9 Example of multilayer coating structure of high barrier film [6]. (SEM courtesy of Vitex Systems, Florida, US) [6]

coated on a film substrate as an electrode, it will provide some barrier function. The typical thickness of an ITO coating is 50–200 nm, and this thickness will provide 10^{-2}–10^{-1} g/m²/day for water vapor transmission and the same order, 10^{-2}–10^{-1} cc/m²/day, for oxygen transmission. Besides an ITO coating, as inorganic layers, SiO_x, Al_2O_3, and SiN_x coatings are often formed using a vapor phase process such as PVD or CVD. Figure 5.9 shows an example of a multilayer coating [6]. In coating them, one needs to be careful about the possible formation of pinholes or microcracks on the inorganic barrier layer caused by dust in the atmosphere or poor handling during the coating process. To avoid degradation induced by such pinhole formation, multiple coatings can provide better barrier performance.

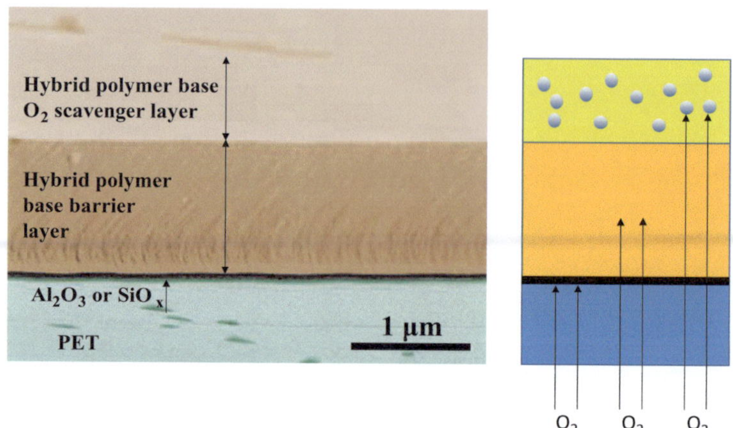

Fig. 5.10 SEM of hybrid barrier layer of passive barrier layer/active oxygen barrier layer and its schematic representation [7]

It has been reported that inorganic–organic polymer hybrid coating provides excellent barrier properties against oxygen and water vapor permeation [5, 7]. These hybrid polymers can be synthesized by a sol-gel technique. The active oxygen barrier layers are produced via sol-gel transformation of a resin consisting of oxidizable organoalkoxysilanes, a UV initiator, and a transition metal catalyst [7]. The barrier film microstructure and the barrier mechanism are schematically shown in Fig. 5.10. With an active barrier layer water vapor transmission can be suppressed by one order or more. The best value reported for water vapor transmission was less than 0.005 g/m^2/day with PET/Al$_2$O$_3$/active hybrid layer and 0.03 g/m$^{2/}$day for PET/Al$_2$O$_3$, while oxygen transmission went from 0.06 cc/m^2/day without an active layer to less than 0.005 cc/m$^{2/}$day with an active layer.

References

1. Nogi M, Kim C, Sugahara T, Inui T, Takahashi T, Suganuma K (2013) High thermal stability of optical transparency in cellulose nanofiber paper. Appl Phys Lett 102:181911
2. Nogi M, Iwamoto S, Nakagaito AN, Yano H (2009) Optically transparent nanofiber paper. Adv Mater 21(16):1595–1598
3. Schaer M, Nüesch F, Berner D, Leo W, Zuppiroli L (2001) Water vapor and oxygen degradation mechanisms in organic light emitting diodes. Adv Func Mater 11(2):116–121
4. Kim SY, Kim KY, Tak Y-H, Lee J-L (2006) Dark spot formation mechanism in organic light emitting diodes. Appl Phys Lett 89:132108
5. Charton C, Schiller N, Fahland M (2006) A. Holla¨nder, A. Wedel, K. Noller, Development of high barrier films on flexible polymer substrates. Thin Solid Films 502:99–103
6. MacDonald WA (2004) Engineered films for display technologies. J Mater Chem 14:4–10
7. Schwab SA, Weber U, Burger A, Nique S, Xalter R (2006) Development of passive and active barrier coatings on the basis of inorganic-organic polymers. Chemical Monthly 137(5):657–666

Chapter 6
Interconnection

6.1 Choice of Interconnection Methods

In electronics manufacturing, an electric interconnection or a system integration step is always required before releasing products. All expected PE products introduced in Chap. 1 require certain kinds of system integration. For instance, displays fabricated using PE technology always require several logic circuits, an energy supply, wireless/wired connection circuits, and other features. Photovoltaic cells require series interconnections and connectors to cables to a battery or a power conditioner. Thus, an interconnection method is one of the essential technologies for PE products. Several important requirements for PE interconnection technology are listed as follows:

- Low temperature: $\leq$80–130 °C
- Task time: less than a few seconds to a few minutes
- Resistance: approx. 10^{-5} to 10^{-4} Ω cm
- Thermal conductivity: as better as possible

The first point is a process temperature. For PE's mass-production market, most products will have low-cost substrates such as PET, PEN, papers, or even acrylonitrile-butadiene-styrene (ABS) resin. The heat-resistance temperatures of PET, papers, and ABS are in the 100–130 °C, 150–200 °C, and 80–100 °C ranges, respectively. Task time must be short if mass production is intended with fast roll-to-roll printing. As for resistivity, even though a lower value is better for electronics, it is expected that the level of current standard interconnection materials such as solders and conductive adhesives will be kept at around 1.2–10×10^{-5} Ω cm. Thermal conductivity will sometimes be required to prevent organic semiconductor devices from heating up since many of them are susceptible to heat and moisture. Of course, requirements will be modified by the characteristics of each PE product.

Figure 6.1 shows typical interconnection methods for electronics with a vertical axis of process temperatures. For many years, interconnection methods have

K. Suganuma, *Introduction to Printed Electronics*, SpringerBriefs in Electrical and Computer Engineering 74, DOI 10.1007/978-1-4614-9625-0_6,
© Springer Science+Business Media New York 2014

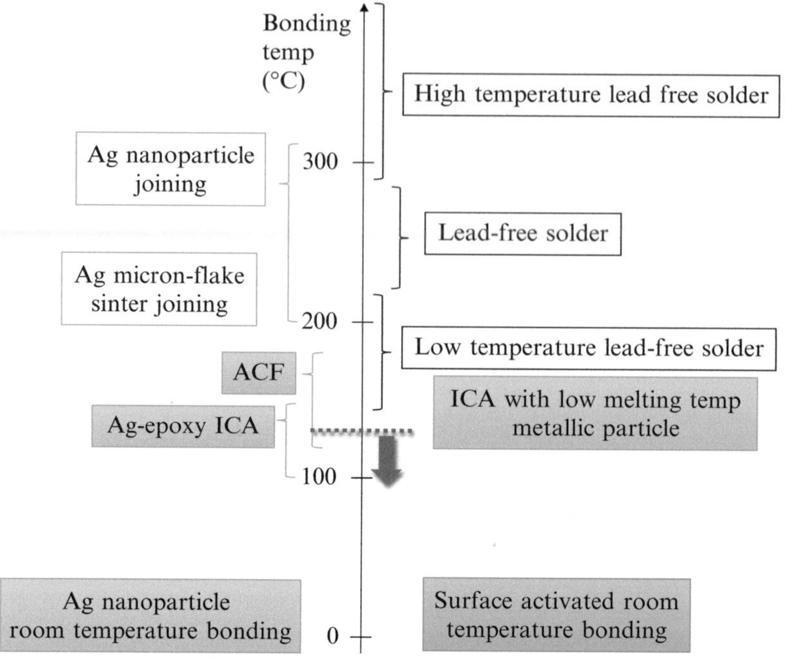

Fig. 6.1 Interconnection methods and their process temperatures

emerged as a result of the revolution in electronics products, and particularly in the last 2 decades, the most commonly used interconnection method, soldering, has undergone a major transition from lead to lead-free soldering [1]. Following this transition, lead-free alloys must be sought out, even for PE technology. Conductive adhesive technology, including both isotropic and anisotropic types, which can have low process temperatures, is attractive for PE products. Nevertheless, the process temperature range required for PE products is far below those of existing low-temperature interconnection materials and processes. In this chapter, variations of interconnection methods and some critical reliability issues are summarized.

6.2   Soldering

The current standard solder is a Sn–Ag–Cu alloy, which has a eutectic temperature of 217 °C. The process temperature, a reflow process, requires more than 240 °C, and it possesses excellent reliability. This alloy, unfortunately, cannot be applied to most PE technology due to its high process temperature. Among solder materials, the Sn–58Bi eutectic solder may be the solder of choice for PE products. It has a low process temperature, 160 °C in the reflow soldering process,

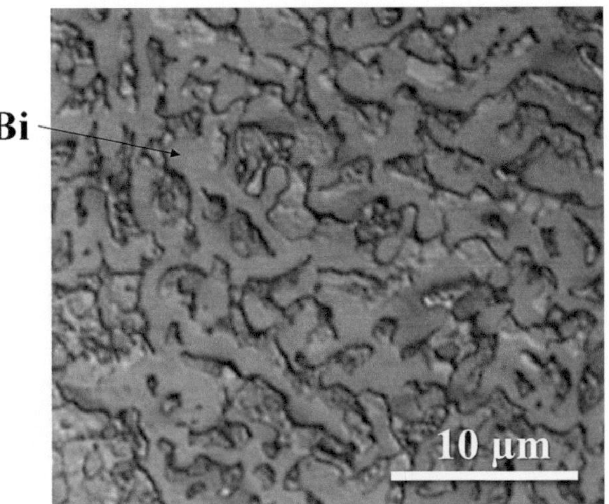

Fig. 6.2 Typical Sn–Bi eutectic microstructure

which is still high for many PE applications. The reliability of the Sn–Bi eutectic solder is excellent when soldered products are used below 100 °C [2]. One of the disadvantages of this solder is its brittleness due to brittle Bi precipitation, as shown in Fig. 6.2. For mobile products, the drop impact may cause damage at the joints easier than other solders. The Sn–52In eutectic alloy has a low melting temperature of 117 °C, and the process temperature may be 130–140 °C. Unfortunately, there little data are available regarding the interconnection of electronic products because the alloy is expensive and a too low heat resistance temperature as a solder interconnection. Its heat resistance is far below 100 °C. The Bi–51In eutectic solder has a very low melting temperature, 89 °C, which could yield a reflow soldering temperature of 100–125 °C [3]. It has been confirmed that the alloy possesses excellent fatigue resistance between −40 and 85 °C, which is close to the melting temperature, due to the dispersion of an intermetallic compound in the alloy matrix.

A soldering process can be modified to lower the soldering temperature or to avoid damage to surrounding components. Local heating facilitates soldering of heat-sensitive PE products. A laser soldering system is shown in Fig. 6.3. The minimum size of a laser beam spot is 50–100 μm, and one can minimize heat damage.

In soldering, one needs to understand the influence of the interface reaction between Sn and the printed electrodes. Both for Ag and Cu electrodes, Sn reacts easily to form thick intermetallic compounds at the interfaces, which are very brittle, which weakens the interfaces. To avoid this severe reaction, electroless Au/Ni plating (ENIG) on electrodes is effective since the Ni–Sn reaction is milder than those with Ag or Cu. During soldering, a thin Au layer is dissolved into the solder immediately on solder melting. Figure 6.4 shows such an example. On an inkjet-printed multilayer circuit film with Ag nanoparticle ink, a Sn-plated chip component was soldered with Au/Ni barrier plating.

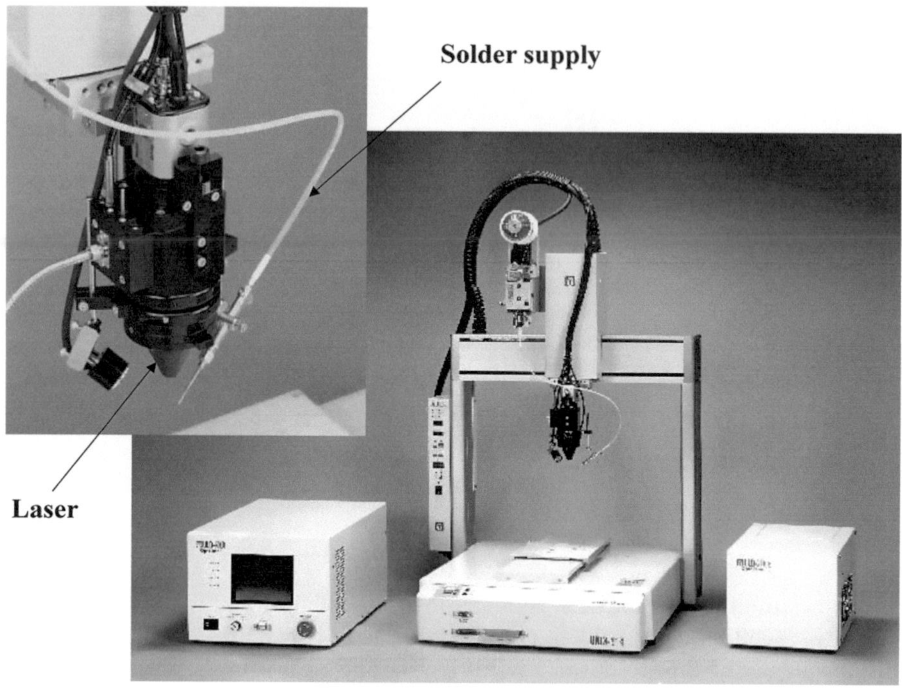

Fig. 6.3 Laser soldering system (Courtesy of Japan Unix, Tokyo, Japan)

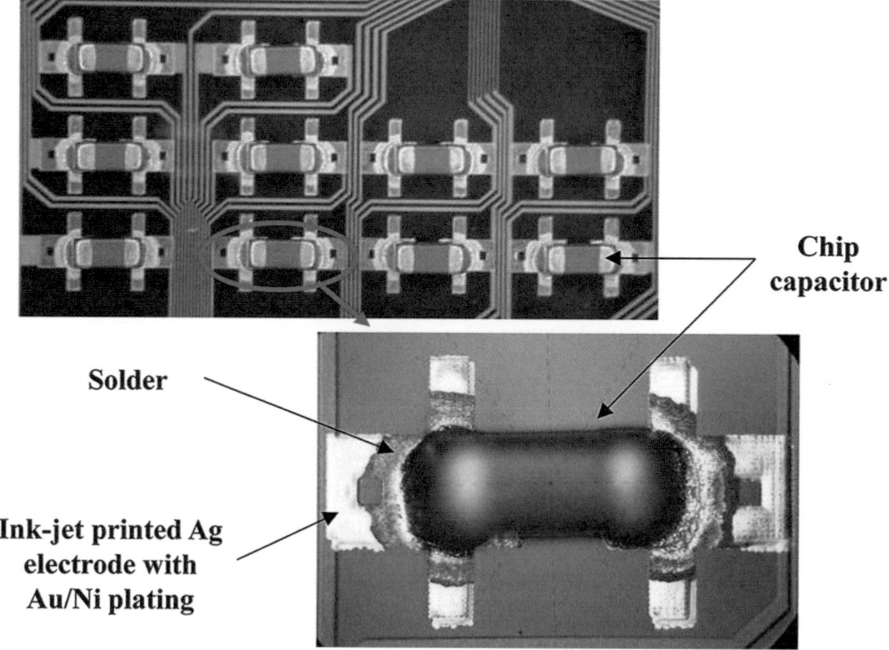

Fig. 6.4 Soldering on Au/Ni plated Ag electrodes (Courtesy of EPSON, Nagano, Japan)

6.3   Conductive Adhesives

Currently, conductive adhesives may be the best choice for PE products. Conductive adhesives have a long history in the electronics industry for more than a half-century. Unlike conductive polymers, discussed in Chap. 3, conductive adhesives are comprised of conductive fillers and a stable adhesive matrix. The stability in air and harsh environments is quite excellent. Today, many forms are available, as listed in Fig. 6.5. The most commonly used matrix adhesive is epoxy because of its high bonding strength, stability in various environments, and low cost. On the other hand, epoxy is rather brittle, and if an application product requires flexibility, epoxy can be modified by incorporating soft segments into its molecular chain. The other method is by using polymer rubbers such as polyurethane or silicone as matrices; these are introduced in the last section. There are two types of conductive adhesive. One is an isotropic conductive adhesive (ICA) that connects in all direction; it can be used as a replacement for solders. The other type of conductive adhesive is anisotropic conductive adhesives; the paste type is called anisotropic conductive paste (ACP) and the film type is called anisotropic conductive film (ACF).

6.3.1   Isotropic Conductive Adhesives

Figure 6.6 shows the variations of commercial ICAs. Screen printing or dispensing is commonly used to print ICAs. Various metallic fillers and carbon fillers have been used as commercial ICAs. Among them, Ag fillers are most often used instead of

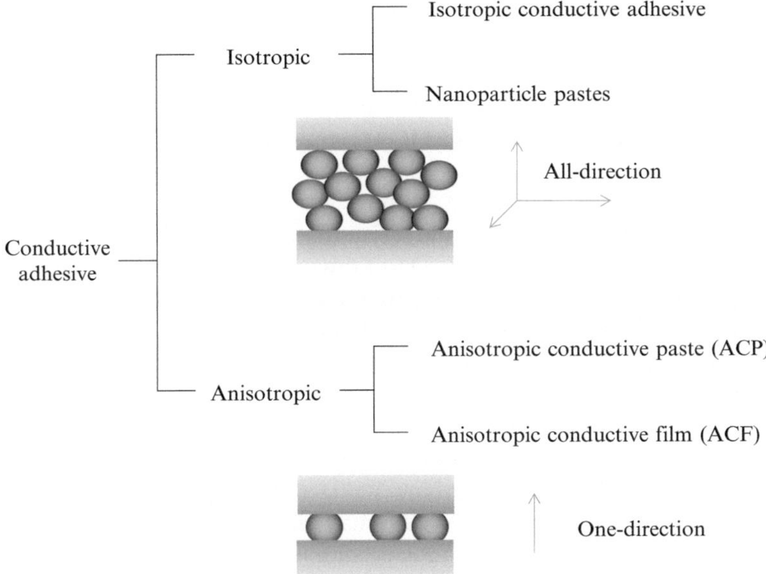

Fig. 6.5 Variation of conductive adhesives

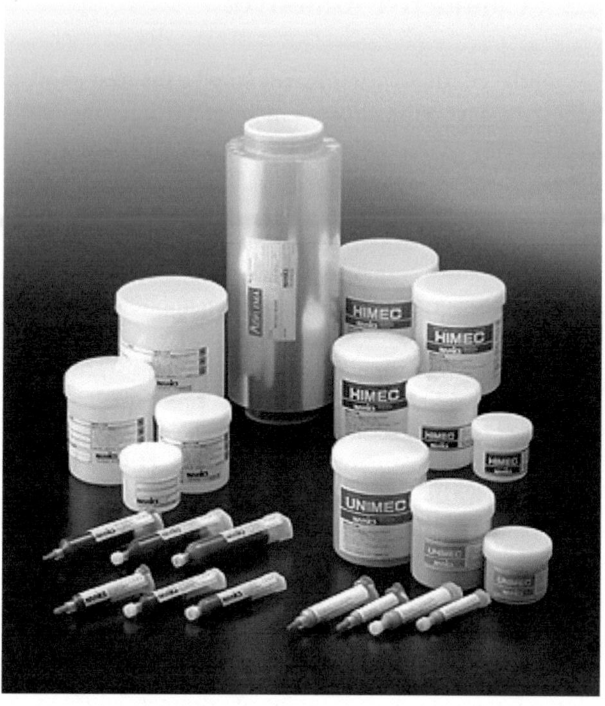

Fig. 6.6 Commercial conductive adhesives (Courtesy of NAMICS, Niigata, Japan)

particulates because of their better connection performance. Figure 6.7 shows the microstructure of Ag metallic flakes, which play a key role in electrical connections, and their typical distribution microstructure in a joint on a circuit board. Ag flakes are well aligned along electrode surfaces, forming good contacts among them. To achieve a good connection, there should be 50–60 vol.% Ag flakes. The resistivity of typical ICAs is in the range of 5×10^{-5} to 1×10^{-4} Ω cm for epoxy matrix and is slightly larger for other matrices. Most ICAs require 150 °C curing due to the nature of epoxy adhesives.

There are, however, a limited number of low-temperature curable ICAs (Table 6.1). As shown in the table, the resistivity is higher by one order of magnitude than those of conventional ICAs. Most low-temperature ICAs possess a lower bonding strength. Even though the temperature requirement is satisfied, curing time is too long for roll-to-roll production. Thus, there is still strong demand for new types of ICAs that possess a lower resistivity, higher strength, and shorter curing time.

ICAs have already been applied to PE technology for many years, as shown in Figs. 6.8, 6.9, and 6.10. Most membrane switches are fabricated by screen printing with ICAs on a PET film. Figure 6.9 shows RFID antennas fabricated by screen printing. As the wiring pitch of touch panels of smart phones and tablets grows finer, current screen printing technology with Ag nanoparticle pastes almost reaches its fineness limit, as shown in Fig. 6.10. The next R&D target will be to come up with a method

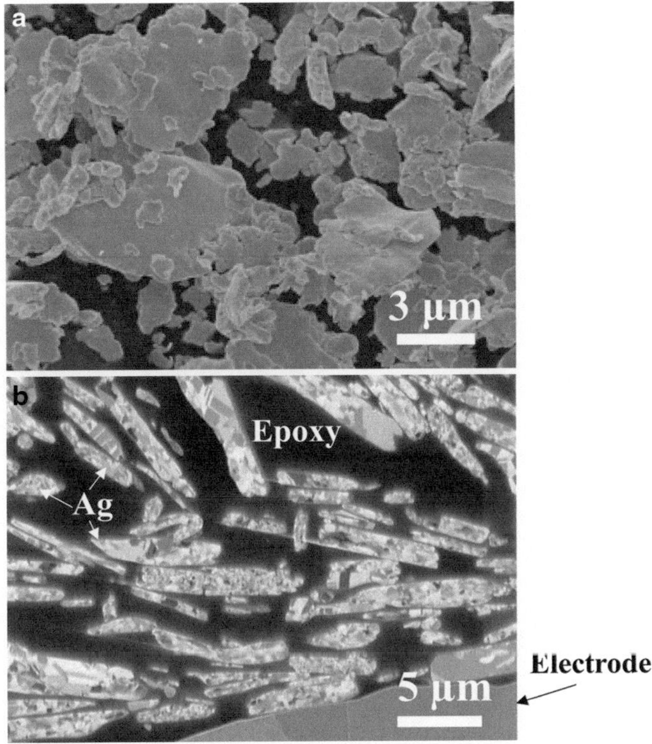

Fig. 6.7 (**a**) Typical Ag flakes and (**b**) their distribution in a bonded conductive adhesive fillet (SEM). White the elongated particles are Ag and the black matrix is epoxy

Table 6.1 Typical commercially available low-temperature curable ICAs

Suppliers	Adhesive	Resistivity (Ω cm)	Bonding strength (MPa)	Curing condition	Storage and notes
A	Epoxy	3×10^{-4}	8	100 °C for 60 min	−20 °C
	Polyester	2×10^{-4}	–	80 °C for 20 min	RT
B	Epoxy	10×10^{-4}	40	80 °C for 120 min	–
C	Epoxy	5×10^{-4}	5.0	80 °C for 60 min	–
D	Modified epoxy	2×10^{-4}	–	120 °C for 90 min 150 °C for 40 min	–
E	Epoxy	5×10^{-4}	6.6	90 °C	RT-30 days
F	Modified silicone	6.5×10^{-4}	1.5	23 °C/50% RH for 11 min	RT
	Epoxy	3.0×10^{-4}	10.6	90 °C for 30 min 120 °C for 10 min	−20 °C
G	Epoxy	10×10^{-4}	15	80 °C for 60 min	Two-part resin

Fig. 6.8 Membrane switch fabricated by screen printing of ICA on PET film

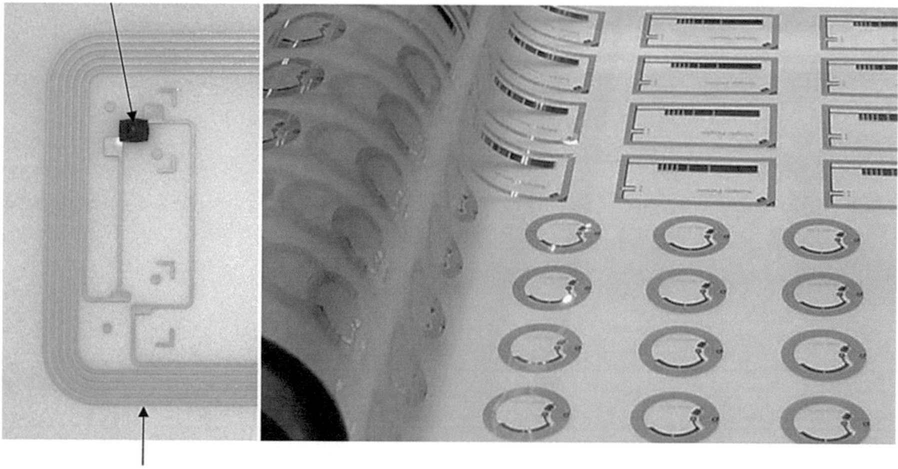

Fig. 6.9 Roll-to-Roll screen printed RFID antennas with ICA

for lowering the resistance value of a picture frame wiring structure to be formed in a narrow frame layer and manufacturing a touch panel with high efficiency.

One of the interesting applications of PE technology is stretchable, not simply flexible, electronics. Stretchable and flexible wiring will create wearable electronics or artificial nerves that can bend on skin and inside the body. ICAs exhibit great potential for such applications. Figure 6.11 shows the yearly progress in the stretchability index, elongation, in the past 1 decade. All stretchable technologies utilize polymer rubbers such as silicone and polyurethane as substrates. There are many types of

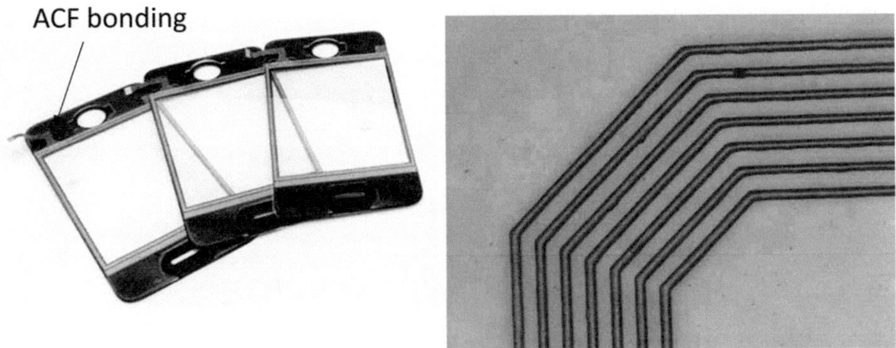

Fig. 6.10 Touch panel of smart phone with fine pitch picture frame wiring. L/S is 30 μm. A flexible cable was bonded with ACF

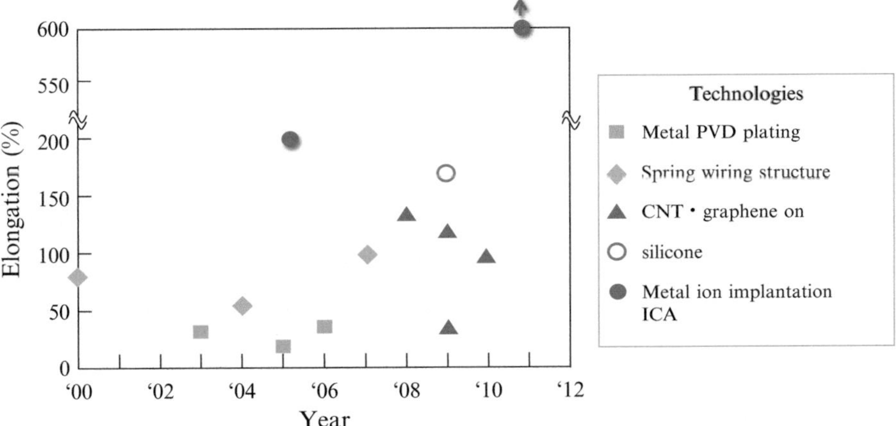

Fig. 6.11 History of stretchable wiring

stretchable wiring technologies, for example, spring patterning or 3D structures, thin-film deposition, metal-ion implantation, CNT/graphene patterning, and ICAs. The author's group was the first to show that Ag flakes with a silicone matrix conductive adhesive exhibited elongation up to 200 % without failure [4]. A pressure sensor floor mat is shown in Fig. 6.12. Dispensing Ag-silicon ICA into a hot oil bath can yield a continuous stretchable conductive string (Fig. 6.13) [5]. The longest elongation was obtained by an ICA consisting of Ag flakes and polyurethane [6]. In this combination, the elongation exceeded 600 %, while conductivity was maintained, as shown in Fig. 6.14. Given the high affinity of polyurethane with silver flakes and substrates, high conductivity was maintained on polyurethane substrates stretched up to a strain of 600 % and on folded cellulose substrates. Thus, since ICAs with rubber matrices, which are basically printable anywhere, possess high stretchability and are affordable, it is expected that ICA applications will expand to wearable electronics.

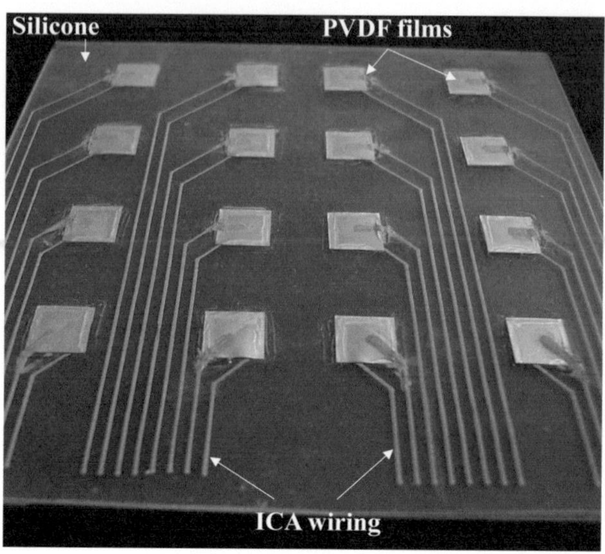

Fig. 6.12 Stretchable floor pressure sensor with PVDF film wiring with Ag-silicone ICA [4]

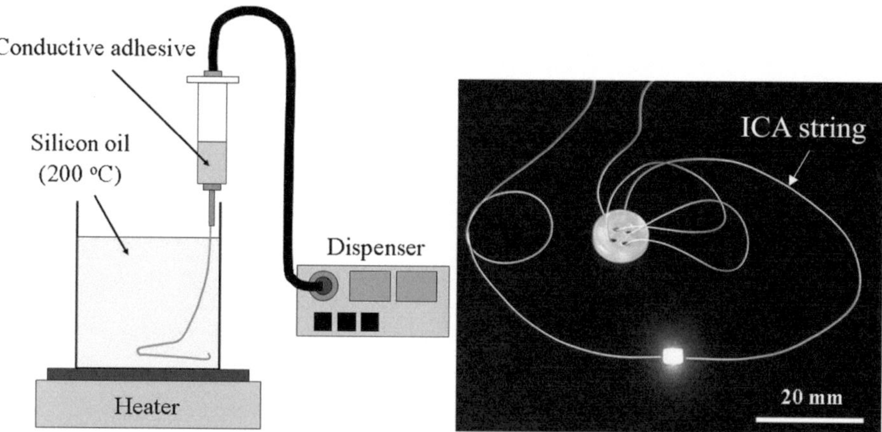

Fig. 6.13 Fabrication of conductive stretchable string with casting into hot oil bath [5]

6.3.2 *Anisotropic Conductive Adhesives*

Unlike ICAs, ACFs and ACPs have been used for fine pitch connections, especially for LCD applications, as shown in Fig. 6.15. ACFs and ACPs can connect at a fine pitch down to a L/S of 10/10 μm, which cannot be achieved with any solders or ICAs. They contain only 2–3 vol.% of conducting particles. Although conductive particles can be metallic balls such as Ni balls with or without a think coating of Au on them, common particles have the structure of a polymer core with a metallic coating. Although metal core balls are used in certain cases, one needs to be careful

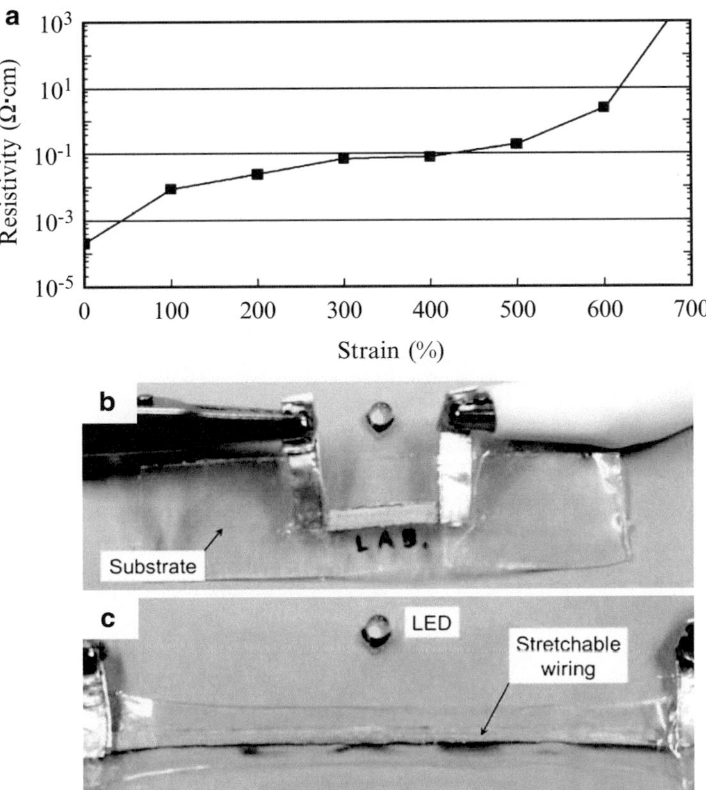

Fig. 6.14 Fourteen stretchable wiring with Ag flakes—polyurethane ICA [6]. (**a**) Resistivity change of polyurethane-based wirings up to a strain of 700 %, (**b**) and (**c**) LED was illuminated with the extremely stretchable polyurethane-based wiring at a strain of 0 and 400 %

with hard metal balls, which may damage devices or substrates during pressing. ACFs are supplied as film reels, as shown in Fig. 6.16. After bonding, balls are compressed as shown in Fig. 6.16. Bonding with ACFs and ACPs is usually carried out using a bonding tool by applying pressure at elevated temperatures. Figure 6.17 shows an example. Bonding temperature is determined by the type of adhesive and is typically 150 °C or higher. Although this high temperature range is little problem for glass or metallic substrates, most PE products with heat-sensitive substrates require lower temperature, less than 130 °C. For ACF/ACP bonding, pressure is always required. Figure 6.18 shows the effect of pressure on the contact resistance between two electrodes on a glass substrate and on a flexible wiring film [7]. To obtain lower contact resistance, more than 300 kPa is required for this case. For the polymer core ball ACF, the contact resistance slightly increases at higher pressure, which is attributed to the rebounding effect of the soft polymer.

Fig. 6.15 ACF used for connecting LCD panel and Si drivers and flexible cables

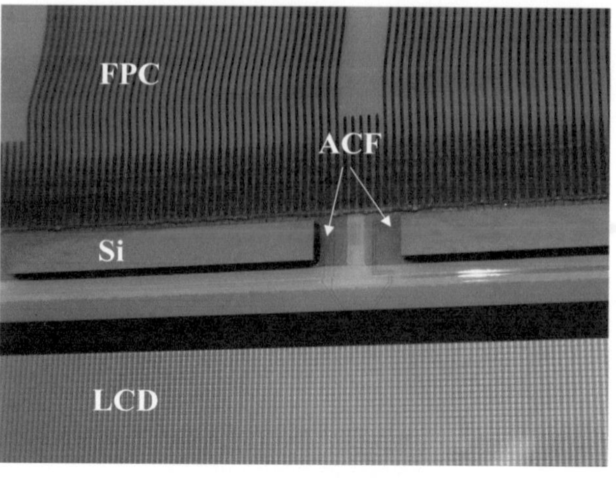

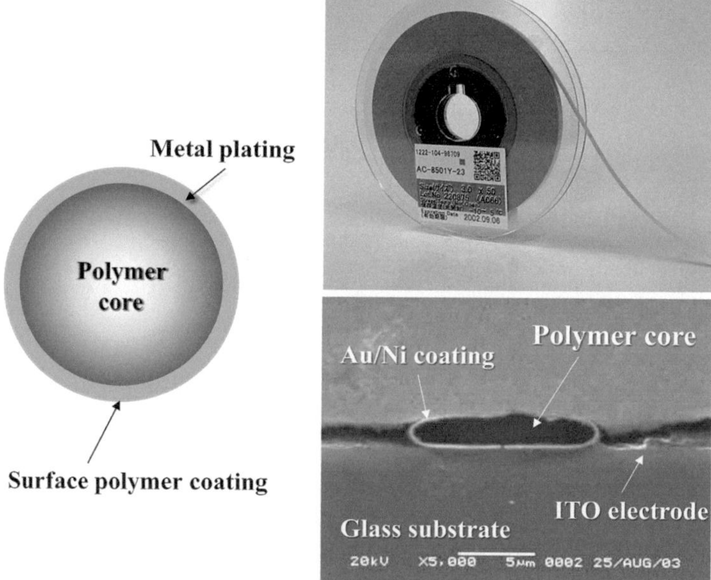

Fig. 6.16 Typical conducting particle structure, ACF roll and bonded microstructure

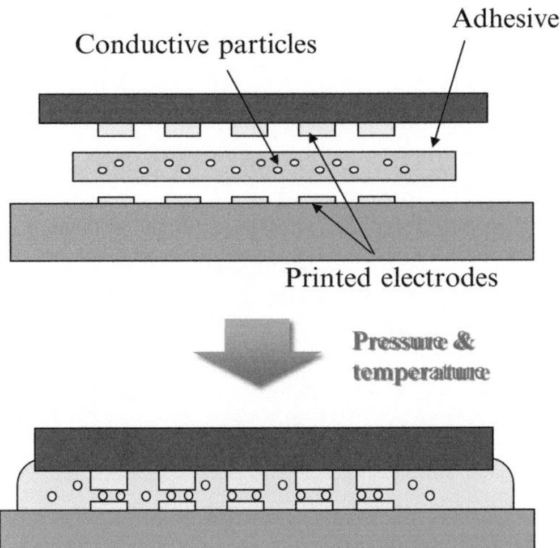

Fig. 6.17 ACF bonding and bonding machine

Fig. 6.18 Effect of bonding pressure and conductive ball materials on contact resistance with ACF on a glass substrate [7]

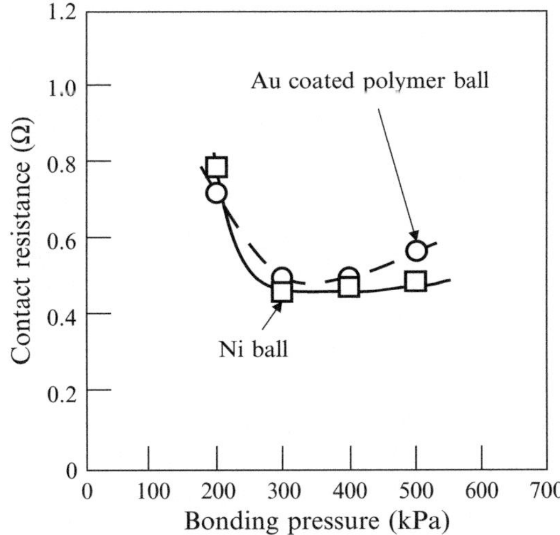

6.4 Interconnection Reliability

To secure the quality and reliability of PE products in the market, one must consider the factors influencing possible product failures in advance and one should reflect those knowledge to PE devise designing, selection of PE materials and of PE manufacturing processes. Many factors must be taken into consideration for a wide variety of PE products, as mentioned in Chap. 1. Products that are completely new, such as a wearable medical devices, about the use of which we have little knowledge, must be explored, and reliability standards must be formulated almost from scratch.

Table 6.2 Thermal cycling conditions and expected lifetime for various electronics products

Products	T_{min} (°C)	T_{max} (°C)	Hold time (h)	Cycles (/year)	Lifetime (year)	Notes
Home appliance	0	60	12	365	10–20	
Note PC/Tablet	–40	85	2	1,825	Approx. 5	5 times on/off
Smart phone	–40	85	12	365	2–5	
Lighting (LED)	0	85	11	365	10–20	
Photovoltaic cell	–40	85	12	365	Approx. 20	
Vehicle (in room)	–55	65	12	365	Approx. 10	
Vehicle (in engine)	–55	150	1	600	Approx. 10	To and from office
Jumbo-jet	–55	95	2	1,200	Approx. 10	Four flights
Fighter	–55	95	2	500	~5	
Satellite	–40	125	1	6,000	5–20	

On the other hand, some existing devices, such as cellular phones or notebook PCs, already have reliability standards. Thus, the present time may be seen as the initial approach to addressing reliability issues for PE technology.

Consider Table 6.2, which table shows typical conditions for various thermal cycling electronic products on thermal cycling. Thermal cycling is one of the major failure modes in current electronics. Mobile devices such as smart phones and notebook PCs or tablets are often subject to harsh environmental conditions due to outdoor use. Devices in a vehicle engine room suffer from the severest temperature conditions, even compared with those for aeronautic applications. In the table, the top six categories are tthe major conditions that occur with PE technology.

Organic electronic semiconductors are quite brittle due to the formation of tight intermolecular packing, which was discussed in Chap. 3. Ceramic devices are also brittle and will suffer from severe thermal expansion mismatch stress when they form on a certain substrate. Table 6.3 compares selected physical properties of PE materials. Ceramic and glass have very small thermal expansion coefficients, less than 1/10, compared to polymers or molecules. Warpage or cracking will occur due to thermal cycling if organic semiconductors are formed on a glass substrate. Even metallic substrates will damage organic/ceramic semiconductors.

Another important reliability issue relates to the effect of humidity on metallic wiring, typically Ag wiring. It is well known that Ag tends to migrate in a humid atmosphere under an electriccal field, which is known as ion migration or chemical migration. Figure 6.19 presents a schematic drawing of ion migration that can occur in typical PE wiring. Ag dendrites grow on circuit boards or in components due to moisture, resulting in short-circuit failures. For given circuits, the tendency of ion migration can be evaluated using very simple tests. Figure 6.20 shows a typical experimental setup for a water drop test, in which a droplet of deionized water is placed between two electrodes. Figure 6.21 shows the effect of applied voltage on the short-circuit failures of fours materials, Cu, Ag-epoxy conductive adhesive, Sn–Ag–Cu, and Sn–Pb solders, measured by a water drop test. As the voltage increases, the time to failure becomes shorter for, in order, Ag-epoxy, Cu, and Sn–Pb.

Table 6.3 Typical physical properties of PE components

Materials	Thermal expansion coefficient ($\times 10^{-6}$/K)	Young's modulus (GPa)	Strength (MPa)	Notes
Pentacene	100–200*	–	–	*TEC of a-axis is negative
ZnO	5*	100–200	–	*TEC of a-axis is 0.4
Si	4	185	2,000–3,000	
SWCNT	–10*	1,000	3,600	*Along longitudinal axis
Graphene	–7 to –1	1,500	130,000	
Ag	18	83	200	
Cu	17	130	190	
Parylene	35–70	2–5	4–7	
PET	60	2–5	55–250	
PEN	13–21	4–12	280–550	
PI	20–60	2.5–3	85–300	
Silicone	270–300	0.005–0.8	0.5–1	
Polyurethane	50–60	0.007–0.07	0.12–7.0	
Glass	4–9	30–90	20–100	
Al	24	70	70–90	
Steel	13	200	400–1,000	
ICA	10–70	5–80	100–150	
Solder	22	50	20	

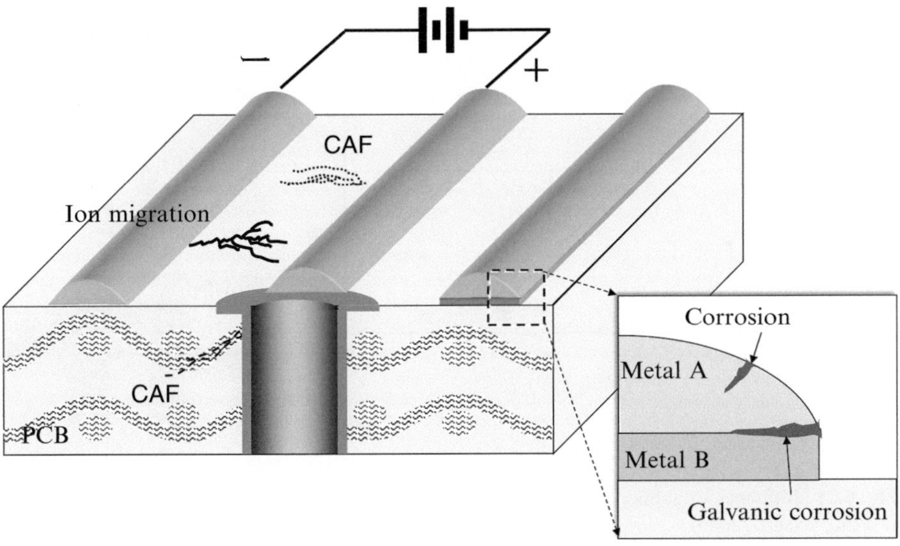

Fig. 6.19 Schematic drawing of possible defects on PCB caused under moisture, ion migration, CAF and galvanic corrosion

Sn–Ag–Cu shows no signs of failure in this voltage range, even though this alloy contains Ag. This result can be attributed to the fact that inside a Sn–Ag–Cu alloy, Ag is fixed as Ag_3Sn, which is a very stable intermetallic compound.

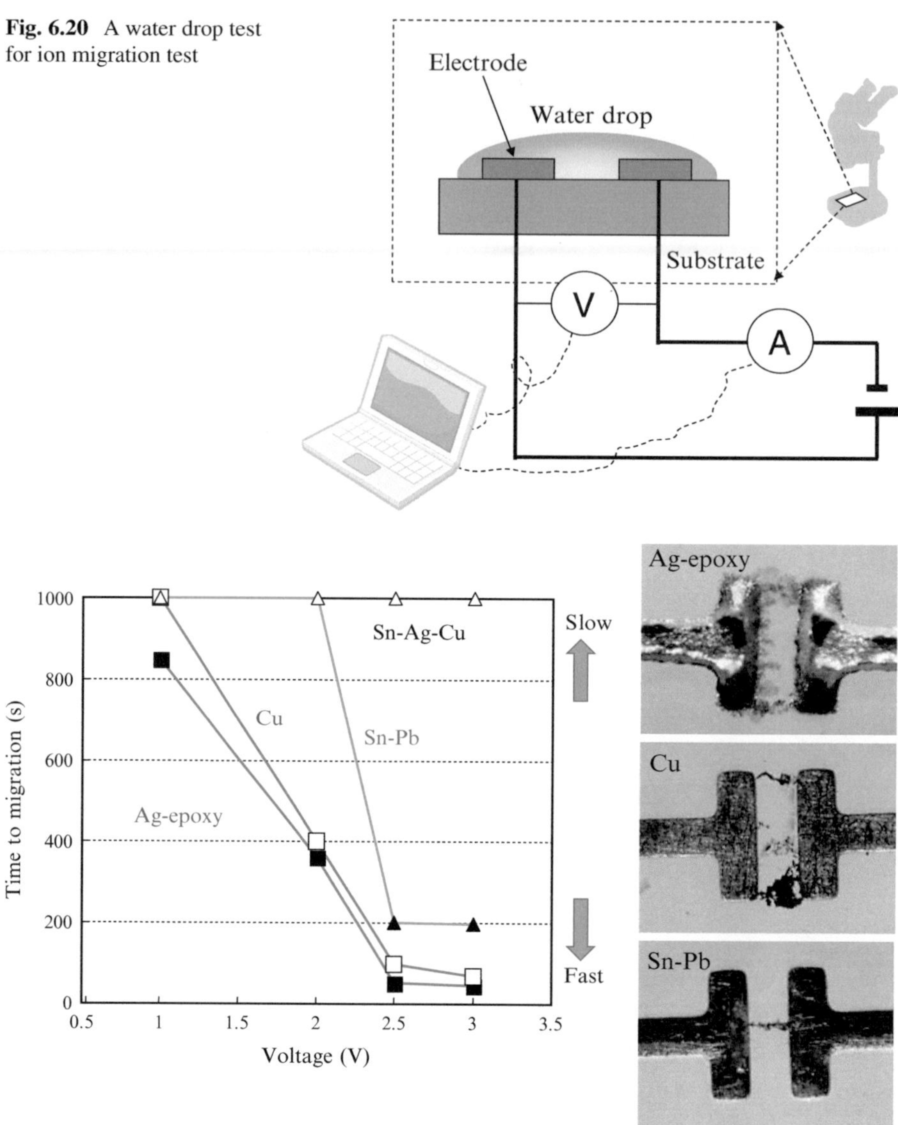

Fig. 6.20 A water drop test for ion migration test

Fig. 6.21 Influence of applied voltage on ion migration failure time (Courtesy of Dr. H. Tanaka, ESPEC, Osaka, Japan)

A water drop test is a simple method to understand a material's tendency toward ion migration. However, this method is not a realistic one for predicting the lifetime of wiring products in a market. For that purpose, a humidity exposure test should be used with a comb-type circuit board defined in the IEC standard [8]. Figure 6.22 shows the experimental setup of this test with comparison data for Ag-epoxy and Cu. Up to 1,000 h, no failure occurred for either Ag-epoxy or Cu wirings. However,

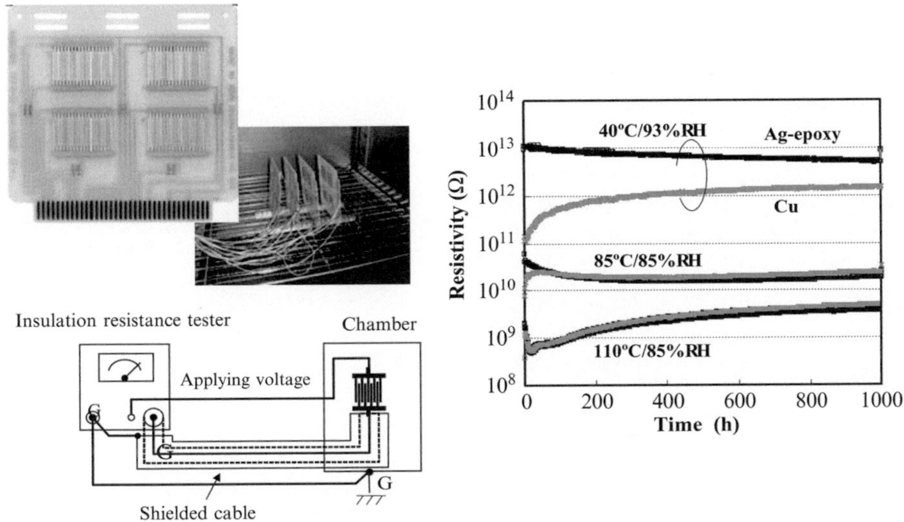

Fig. 6.22 Humidity exposure test with comb type test circuit boards and resistivity change comparison of Ag-epoxy conductive adhesive and Cu (Courtesy of Dr. H. Tanaka, ESPEC)

by a careful observation of tested circuit boards, Ag migration was found, as shown in Fig. 6.23. At 85 °C/85%RH and at 110 °C/85%RH, Ag dendrites grew over a distance of approx. 100 μm, which cannot produce a short-circuit failure because the electrode space is much wider. In addition, it is noteworthy that Cl is always detected at the roots of Ag dendrites, which implies a strong influence of Cl on Ag migration. Cl is usually contained in epoxy as an impurity up to 1,000 ppm.

Another electrochemical corrosion failure can occur at a contact where different metals contact each other, even without externally applied voltage. Figure 6.24 presents this corrosion mechanism schematically. Metals and alloys have their own electrode potentials. When two metals come into contact in an electrolyte such as water, a precious metal acts as cathode and the other base one as anode. The electropotential difference between these dissimilar metals is the driving force for an accelerated corrosion on the anode base metal. The base metal atoms are ionized into metal ions, which end up dissolving into the electrolyte, and the electrons migrate to the precious metal through the interface. The base metal ions are deposited on the cathode metal. Figure 6.25 shows an example of the interface resistance change in high humidity–high temperature exposure for an Ag/Sn interface as compared with an Ag/Au interface [9]. While no increase in resistivity was observed for the Ag/Au interface, which has little electropotential difference, the Ag/Sn interface exhibited a rapid increase in contact resistance up to 300 h. Figure 6.26 shows the interface microstructure change in the humidity/temperature exposure [9]. After severe humidity/temperature exposure, the interface microstructure undergoes a drastic change. As seen in Fig. 6.26b, a thick Sn oxide layer is formed on the Sn surface inhomogeneously, which is clearly identifiable in the TEM photograph of Fig. 6.26c.

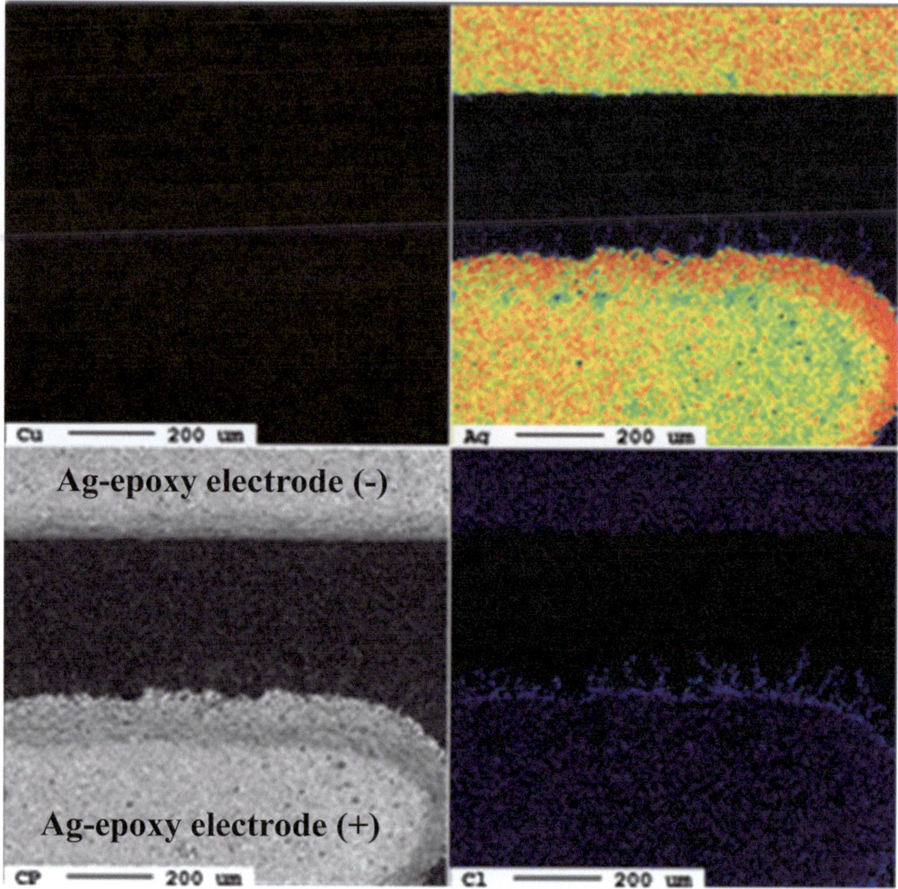

Fig. 6.23 EPMA of Ag ion migration with Cl exposed at 85 °C/85%RH for 1,000 h

Interestingly, a wavelike band is found inside the epoxy matrix along the first array of Ag particles, as indicated by thick arrows in the picture. This wavelike band was found only after a few tens of hours of exposure and did not move deep inside the conductive adhesive. High-resolution imaging revealed that this wave band was an agglomeration of nano SnO particles in an epoxy matrix. In addition, there were many SnO particles dispersed between the wave band and the SnO layer on the Sn plating. Thus, this degradation was caused by Galvanic corrosion as follows:

1. In humidity exposure, H_2O is absorbed into an epoxy matrix, resulting in the formation of a conductive atmosphere at the contacts of Ag particles/Sn surface.
2. Due to the electrochemical potential difference between Ag and Sn, Sn, the base side, is ionized into Sn^{4+}, resulting in dissolution into a conductive epoxy matrix.
3. Due to the potential gradient between the Sn surface and the first array of Ag particles, Sn^{4+} migrates toward the first line of Ag particles. Sn^{4+} ions do not move forward deep into the conductive adhesive since there is no electrochemi-

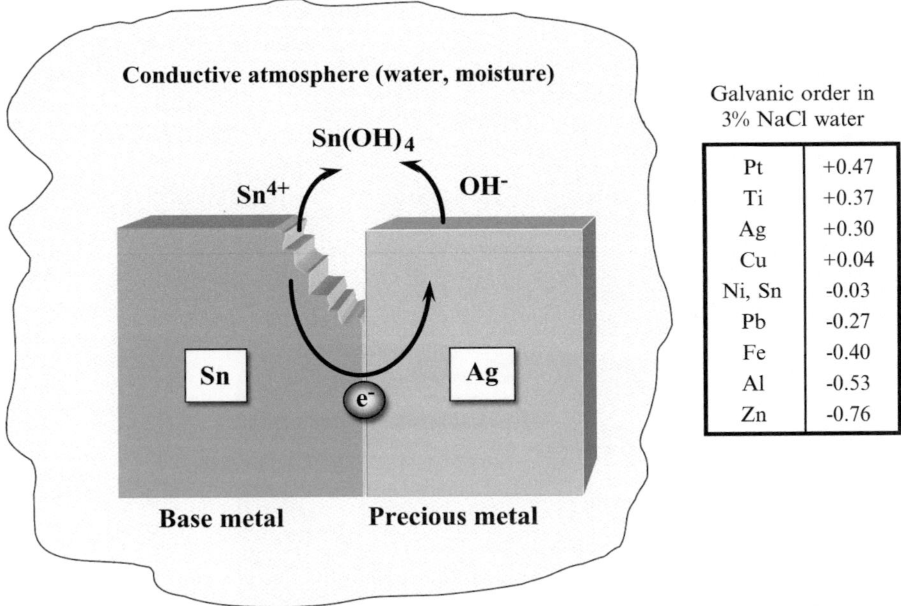

Pt	+0.47
Ti	+0.37
Ag	+0.30
Cu	+0.04
Ni, Sn	-0.03
Pb	-0.27
Fe	-0.40
Al	-0.53
Zn	-0.76

Fig. 6.24 Electrochemical reaction corrosion: galvanic corrosion

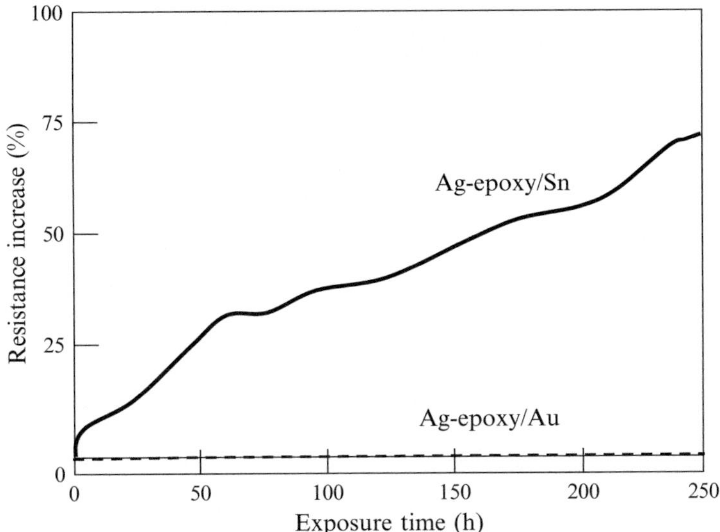

Fig. 6.25 Interface resistance change as a function of exposure time at 85C/85%RH [9]

cal potential difference beyond the first array of Ag particles inside the conductive adhesive.

4. Sn^{4+} forms a Sn oxide at the interface and nanoparticles in an epoxy matrix at elevated temperature.

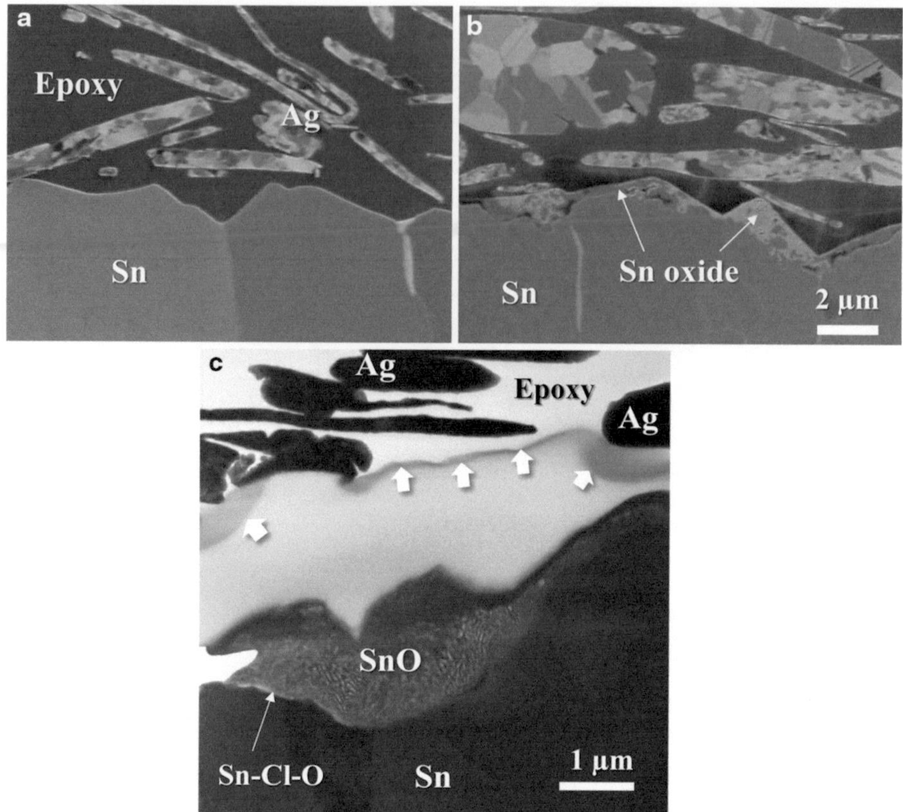

Fig. 6.26 Microstructure of galvanic corrosion in Ag-epoxy/Sn contact exposed at 85 °C/85%RH for 300 h [9]. (**a**) As-received interface and (**b**) after exposure for 300 h (SIM). (**c**) TEM of (**c**)

As indicated in Fig. 6.26c, there is always Cl at the interface; thus, it is likely that Cl acts as a certain kind of catalyst to enhance the ionization of Sn. This was also observed in Ag migration (Fig. 6.23). From these observations, epoxy that is free of Cl or halogen is strongly recommended to avoid moisture-induced corrosion degradation [10].

Interface resistivity is also strongly influenced by moisture in the atmosphere. Figure 6.27 shows the changes in the interface resistance between Ag-epoxy conductive adhesive and a Cu electrode cured under various conditions [11]. Since Ag and Cu do not have a large electrochemical potential difference, the degradation is not caused by Galvanic corrosion. While a fully cured sample, cured at 140 °C for 10 min, shows no significant degradation, samples cured at a lower temperature or for a shorter time exhibit apparent degradation with increasing exposure time. The degradation can be simply attributed to the surface oxidation of a Cu electrode by the raw-cured epoxy's absorption of moisture.

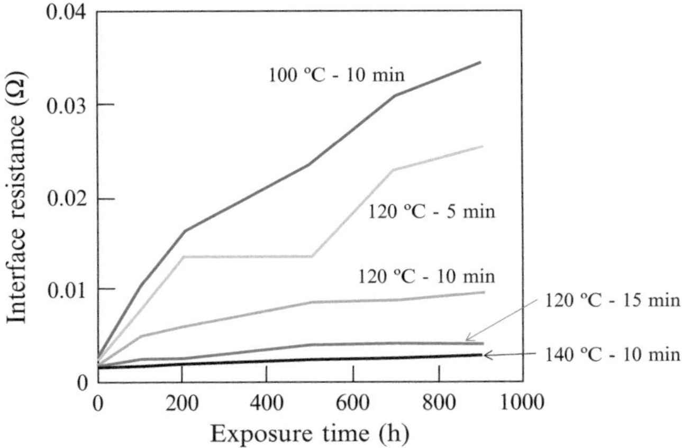

Fig. 6.27 Influence of curing conditions of epoxy on contact resistance between Ag-epoxy conductive adhesive and Cu under humidity exposure at 85 °C/85%RH [12]

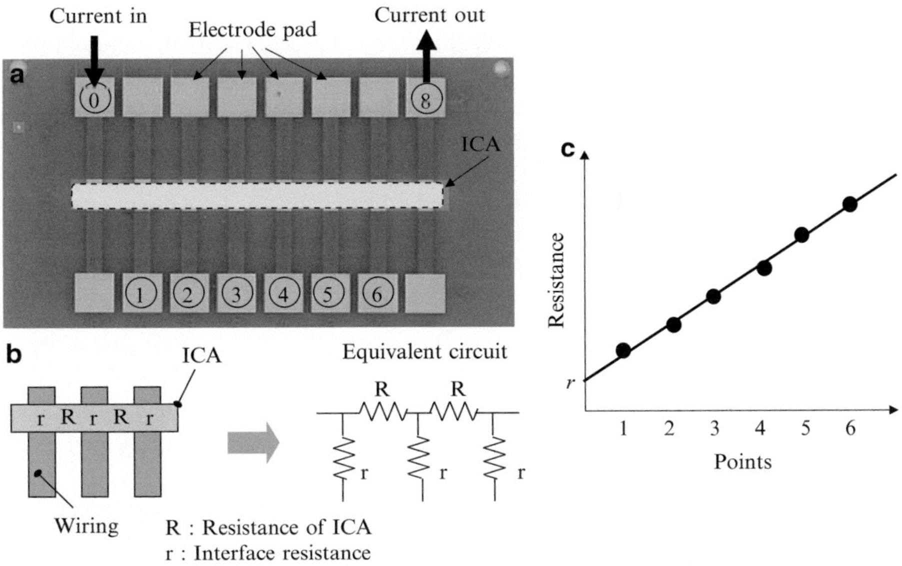

Fig. 6.28 Interface resistivity measurement method defined by ISO [12]. (**a**) A test board, (**b**) a schematics of wiring and its equivalent circuit, and (**c**) a data plot example. The test board is simplified to understand the measurement mechanism, which is slightly different from the ISO definition. By plotting measured voltage at as shown in the graph on the right, the extrapolation to the y-axis provides a y-intercept equivalent to the interface resistance "r"

Thus, in addition to the influence of moisture on plastic substrates/barrier films mentioned in the previous chapter, interface degradation in moisture is a critical issue in PE technology. Recently, the measurement of the interface resistance change in a given environment was established by ISO and will be released in late 2013 [12]. Figure 6.28 illustrates the measurement method, which is similar to

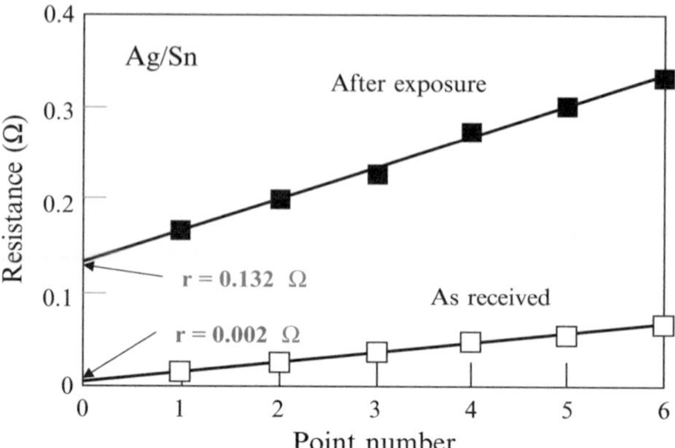

Fig. 6.29 Contact resistivity change measured by method defined in ISO-16525-2. The sample was exposed to 85 °C/85%RH for 1,000 h. By this test condition, Ag/Cu or Ag/Au interfaces did not show substantial degradation

the so-called TLM method, a contact resistance measurement for semiconductors. In this test board, eight stripes are vertically aligned parallel to each other. In the centers of the stripes, the horizontal windows are open so that they can make contact with the Ag-epoxy conductive adhesive line, printed as shown in the picture. From electrode ⓪ to electrode ⑧, a constant current is applied. The voltages at point ①–⑥ are plotted as shown in Fig. 6.28c. The extrapolation to the y-axis provides "r", which is the interface resistance as shown in the equivalent circuit in Fig. 6.28b. Figure 6.29 shows an example of Ag-epoxy/Sn interface resistance measured using this method. Thus, since this evaluation method can detect a slight change in interface resistivity in addition to simplicity, it will provide useful information on contact resistance for a wide variety of interfaces in PE technology.

References

1. Suganuma K (2001) Advances in lead-free electronics soldering. Curr Opin Solid State Mater Sci 5(1):55–64
2. Suganuma K, Sakai T, Kim K-S, Takagi Y, Sugimoto J, Ueshima M (2002) Thermal and mechanical stability of soldering QFP with Sn-Bi-Ag lead-free alloy. IEEE Trans Electron Packag Manuf 25(4):257–261
3. Suh D, Hwang C-W, Ueshima M, Sugimoto J (2008) A novel low-temperature solder, based on intermetallic-compound phases: alloy design, high-homologous temperature properties, and reliability. JOM 62(12–13):71–76
4. Inoue M, Kawahito Y, Tada Y, Hondo T, Kawasaki T, Suganuma K, Ishiguro H (2008) A super-flexible sensor system for humanoid robots and related applications. J Jpn Inst Electron Packag 11(2):136–140
5. Wakuda D, Suganuma K (2011) Stretchable fine fiber with high conductivity fabricated by injection forming. Appl Phys Lett 98:073304
6. Araki T, Nogi M, Suganuma K, Kogure M, Kirihara O (2011) Printable and stretchable conductive wirings comprising silver flakes and elastomers. IEEE Electron Device Lett 32(10):1424–1426

7. Yim M-J, Paik K-W (1998) Design and understanding of anisotropic conductive films (ACF's) for LCD packaging. IEEE Trans Compon Packag Manuf Technol 21(2):226–234
8. IEC 60068-2-67, Environmental testing—Part 2: Tests—Test Cy: Damp heat, steady state, accelerated test primarily intended for components (1995).
9. Kim S-S, Kim K-S, Lee K, Kim S, Suganuma K, Tanaka H (2011) Electrical resistance and microstructural changes of silver-epoxy isotropic conductive adhesive joints under high humidity and heat. J Electron Mater 40(2):232–238
10. Kim K-S, Lee K, Suganuma K, Huh S-H (2011) Effect of Cl content on interface characteristics of isotropic conductive adhesives/Sn plating interface. J Microelectron Packag Soc 18(3):33–37
11. Klosterman D, Li L, Morris J (1998) Materials characterization, conduction development, and curing effects on reliability of isotropically conductive adhesives. IEEE Trans Compon Packag Manuf Technol 21(1):23–31
12. ISO 16525-2 Adhesives—Test methods for electrically isotropic conductive adhesives—Electric characteristic for electronic assemblies (DC test) (2013).

Chapter 7
Next Step

This book focused on the key technologies of PE, including expectations for the future. Development is under way, and it is reflected in the rapid evolvement of organic and oxide semiconductors even though the technology has just begun. It is true that this field is growing with the revolution in new materials technology. In other words, it is possible that more very new materials will be discovered next year—or even tomorrow. Although some may profit from the discovery of new materials, timing the release of products made with those materials may prove difficult. At any rate, the transition from conventional electronics manufacturing to the new printing fields or new printed products, although no one can say which factor becomes the priority, has begun. It may take a little more time to develop, on a large scale, the expected huge markets such as displays, photovoltaics, and lighting with printable OLED technology.

There are many interesting issues that remain to be experimentally and theoretically addressed in this field. Clearly, the possibility of investigating the intrinsic limits of organic and oxide semiconductors will continue to attract research interest in those new materials, further expanding our understanding of the fundamental electronic properties of printed semiconductors. Wiring technology is also of interest. Room-temperature wiring is made possible using either the new ink or postprocessing. There is no doubt that new TCFs, which will replace the conventional ITO, will play a key role in opening the huge PE market. An important lesson to be drawn from the past is that such a quest must be conducted with great care and scientific rigor, with results subjected to independent verification and reproducibility.

Here, in this final chapter, the activity in the field of PE technology standardization is briefly introduced. Standardization usually happens when a technology has reached a certain point in its development, resulting in a certain level of mass production in the market. PE technology has apparently not yet reached this level. The primary aim in the standardization of PE is to help open new markets by first establishing a common platform for PE technology.

K. Suganuma, *Introduction to Printed Electronics*, SpringerBriefs in Electrical and Computer Engineering 74, DOI 10.1007/978-1-4614-9625-0_7,
© Springer Science+Business Media New York 2014

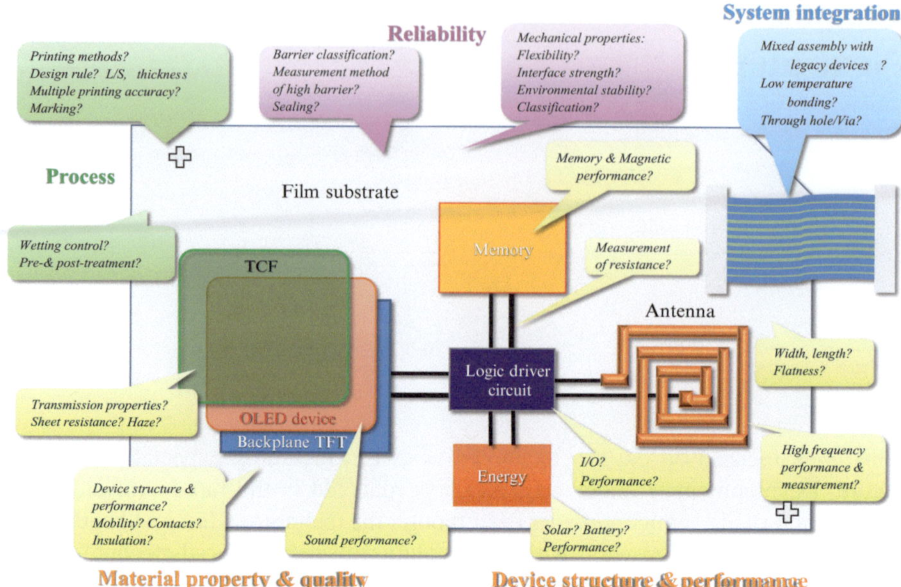

Fig. 7.1 Influential factors in PE technology

Let us consider the possible factors, even though the targets are widely dispersed, spurring growth in PE. Take a look at Fig. 7.1, which depicts the possible factors affecting every component on a flexible film substrate. The standardization issues will be categorized into five groups, i.e., material property and quality, device structure and performance, process, system integration, and reliability. For instance, one can select an organic semiconductor with a certain performance, which is one of the material key properties. Nevertheless, the performance can be easily changed by the device size and structure because charge carriers travel only short distances in organic semiconductors. Designers of devices must make a narrow source-drain gap to obtain the best performance. Nevertheless, we have no such. The surface of the substrate should be smooth enough to obtain a uniform layer structure of electrodes and of light-emitting diodes/photovoltaics. But, we have no guideline. Thus, there are so many factors to consider in the fabrication of a given product. At this moment, in late 2013, no database of materials/substrates/printing has been developed, and manufacturers must start their processes with the selection of materials using a suitable printing method and a performance/reliability evaluation method and at the same time design some target device. Clearly, it will take a considerable amount of time to create an environment that is mature enough for PE mass production.

To establish a common platform for a PE industry supply chain, much effort has been devoted to creating standards for PE technology from several organizations. The most active and advanced organizations are the International Organization for Standardization (ISO), International Electrotechnical Commission (IEC), and IPC. Other organizations are also active, such as Institute of Electrical and Electronics Engineers (IEEE) and Semiconductor Equipment and Materials International (SEMI).

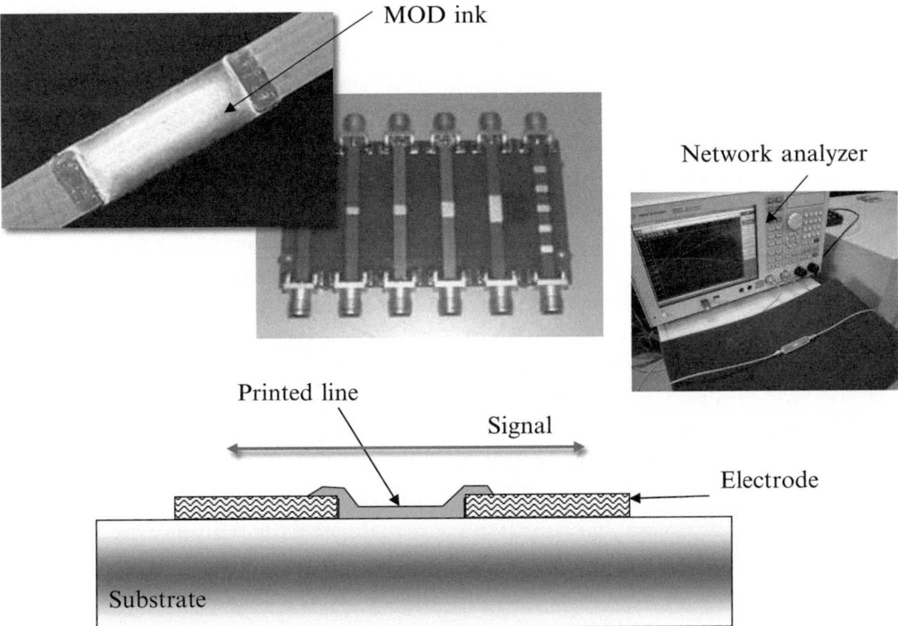

Fig. 7.2 High-speed signal transmission characteristic test methods defined in ISO 16525 "Adhesives—Test methods for isotropic electrically conductive adhesives—Part 9"

ISO has prepared a series of standards for the evaluation of conductive adhesives in ISO/TC 61/WG5/SC11. It will be published as ISO 16525: "Adhesives—Test methods for electrically isotropic conductive adhesives" [1] will consist of nine parts, as follows:

Part 1: General test methods
Part 2: Electric characteristic test methods for electronic assemblies
Part 3: Heat transfer properties
Part 4: Shear strength test method for rigid to rigid bonded assembly
Part 5: Shear fatigue test method
Part 6: Pendulum type shear impact test method
Part 7: Environmental test methods
Part 8: Electrochemical migration test methods
Part 9: High speed signal transmission characteristic test methods

Of these, Parts 2, 7, 8, and 9 are closely related to PE technology. Some of Parts 2, 7, and 8 were already introduced in Chap. 6. In Part 9, high-speed transmission measurement methods are described in detail. High-speed transmission, which relates to high-frequency devices, is one of the essential properties of RFIDs. This part contains the measurement method both for bonding interconnects and for printed wiring. Figure 7.2 shows a printed circuit board (PCB) for the high-frequency transmission measurement of printed wiring. The test itself is simple, but noise cancelation in the measurements should be carefully performed. An outline of time domain reflection (TDR) measurement is as follows: connect one end of the transmission line to be measured to the oscilloscope, leaving the other end unconnected; choose the TDR mode of the oscilloscope, with the impedance indicated on the y-axis, and record the result. Since an electrically isotropic conductive adhesive will be applied to the center of each standard test circuit board, set the marker at the midpoint between the input port with a SMA profile (connector, subminiature Type A) and the output port with a SMA profile. Obtain the value of the characteristic impedance at the marker position indicated. Using this method, a high signal transmission greater than 10 Gb/s can be measured. Including all the other parts, this will be published in early 2014.

Another ISO activity involves the standardization of a high barrier evaluation method, which also falls under TC 61/SC 11. Compared with glass substrates, a plastic substrate has poor water vapor and oxygen barrier properties. The requirement for high barrier films and sealants used for OLED devices is quite strict, 10^{-4} to 10^{-6} $g/cm^2/day$, as mentioned in Chap. 5. Although there are several measurement methods, the permeation rate is much lower than the analytical limit of current commercial technologies. So far, various high barrier analysis methods have been proposed, but they are still in the developmental stage. New standards—ISO 15106-5, 15106-6, and 15106-7—are intended to provide industrial tools for measuring high barrier properties [1].

IEC Technical Committee (TC) 119, which is devoted to PE, was launched in 2012 [2]. The role of TC 119 is to facilitate the rollout and industrialization of PE technology. The mission of TC119 is formally setting down the standardized terminology, materials, printability, and devices by printing methods applicable worldwide [3]. TC 119 has active representations from ten countries: China, Germany, UK, Italy, Finland, Japan, South Korea, Russia, Sweden, and the USA, with a further eight countries taking part as observers. TC 119 has already established six ad hoc working groups to promote this process, as shown in Fig. 7.3.

Usually, ISO and IEC standardization procedures require 2 or 3 years to establish a single standard. Such long periods are sometimes too long for fast-growing industries. IPC, the American printed circuit board association, and its members are actively engaging in global printed electronics efforts dedicated to furthering the competitive excellence and financial success of IPC members. IPC's Printed Electronics Initiative was established in 2011 to lead element required for developing printed electronics industry worldwide. This initiative has already resulted in the first two operational-level standards in printed electronics, followed by one draft, as shown in Fig. 7.4 [4].

IPC/JPCA-4921: Requirements for Printed Electronics Base Materials

IPC/JPCA-4591: Requirements for Printed Electronics Functional Conductive Materials

IPC/JPCA-2291: Design Guidelines for Printed Electronics is scheduled for publication later this year

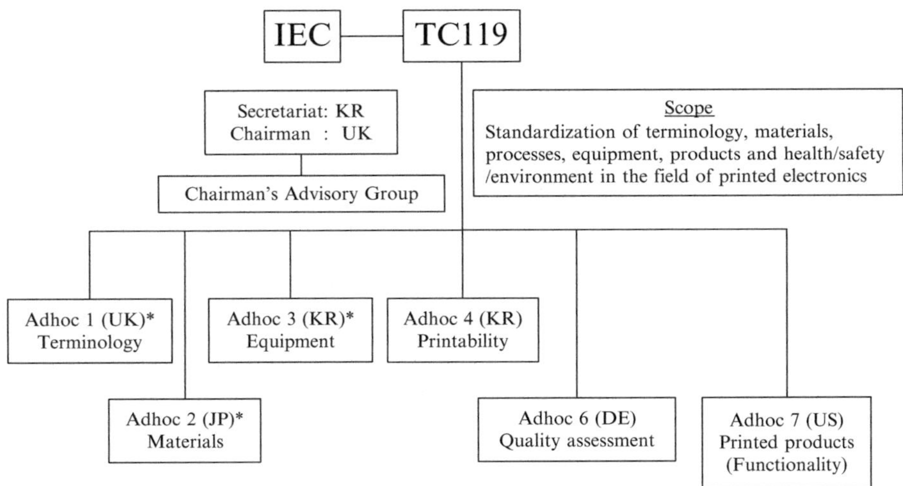

Fig. 7.3 TC119 PE organization structure

Currently, standards development committees are working on printed electronics standards, as shown in Fig. 7.4. They emphasize collaborative efforts with other organizations to speed developments, as evidenced by the joint standards codeveloped with JPCA colleagues. Cooperative efforts with the IEC, NPES (Association for Suppliers of Printing, Publishing and Converting Technologies), iNEMI, and others are also under way.

Finally, it should be repeated that the PE industry is currently in development. One should watch the revolution very closely because many new materials and processes related to PE technology are emerging all the time. At the same time, when a new technology is trying to replace legacy technology, there will be always competition between them. Sometimes, the legacy technology wins out. This has happening in the PE world as well.

Photovoltaics with organic semiconductors have been one of the hottest PE technology fields, as discussed in Chap. 1. It is believe that photovoltaics will be the leading technology in renewable energy. Organic photovoltaics are competing with the legacy Si technology due to unexpected cost reduction of Si solars. This situation is very similar to the earlier DRAM or LCD panel supply chains problem. About 2 decades ago, when the manufacture of those digital devices required the most advanced electronics technology, a high level of research capability supported activity in those cutting edge fields. Nevertheless, the technology turned out to be strongly dependent on facilities. When an industry matures, those with the resources to purchase the manufacturing facilities are able to compete. At that point, prices start to fall sharply. This is happening again in PE. For instance, the most advanced organic photovoltaic thin-film manufacturer, Konarka Technologies, filed for bankruptcy in 2011. The other example concerns e-paper technology versus LCDs. No one could have predicted such an abrupt reduction in costs in the LCD market. A

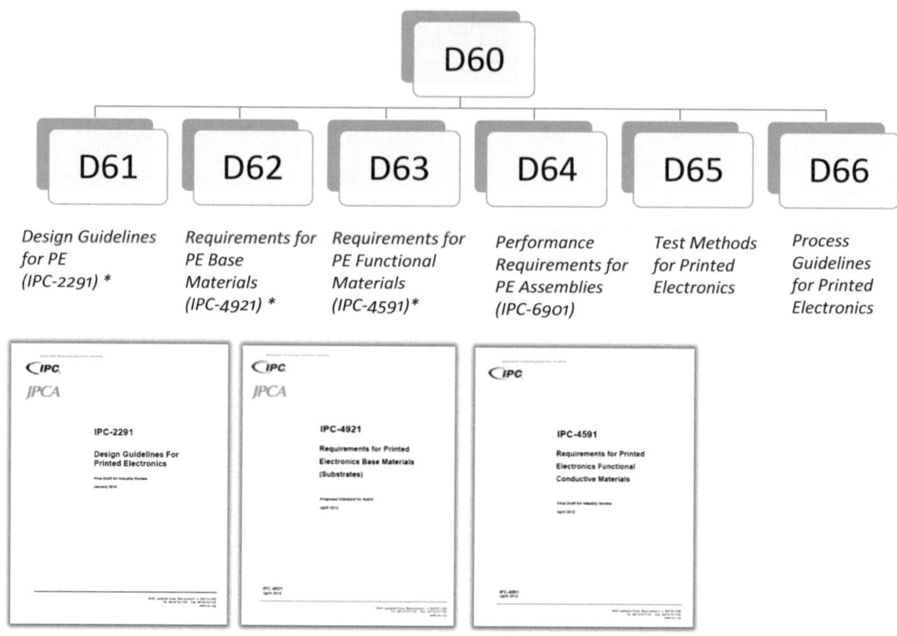

Fig. 7.4 IPC PE initiative

flexible and rollable display was and is one of the most attractive technologies for wearable devices. Nevertheless, the commercialization of printed devices with rollable displays or printed organic TFT panels was interrupted by the sudden arrival of smartphones and tablets equipped with brilliant LCDs and possessing fine-scale functions. Thus companies that are developing printed electronic devices should be aware that they are competing with the legacy device technology and its cost-cutting trends.

Flexibility and ultimate cost reductions could be the most promising features of PE. The competition, not only within the PE industries but also with the legacy industry, is cutthroat. With this fact in mind, readers and consumer should look forward to a bright future in the field of PE.

References

1. ISO TC61/WG5/SC11. http://www.iso.org/iso/home/store/catalogue_tc/catalogue_tc_browse.htm?commid=49424
2. IEC TC119 Printed Electronics. http://www.iec.ch/dyn/www/f?p=103:7:0::::FSP_ORG_ID,FSP_LANG_ID:8679,25
3. Lee H (2013) Industrialization of printed electronics via standardization—IEC TC119. Presented at ICFPE2013, Jeju, Korea, September 11–13
4. IPC Printed Electronics Initiative. http://www.ipc.org/ContentPage.aspx?pageid=Printed-Electronics-Initiative